Ideal Platforms for Optimizing IT Workloads
TRIED AND TESTED SOLUTIONS TO ACCELERATE IT AND BUSINESS TRANSFORMATION

First Edition

Marty Poniatowski

HPE Press
660 4th Street, #802
San Francisco, CA 94107

Ideal Platforms for Optimizing IT Workloads
Tried and tested solutions to accelerate IT and business transformation

Marty Poniatowski

© 2017 Hewlett Packard Enterprise Development LP.

Published by:

Hewlett Packard Enterprise Press
660 4th Street, #802
San Francisco, CA 94107

All rights reserved. No part of this book may be reproduced or transmitted in any form or by any means, electronic or mechanical, including photocopying, recording, or by any information storage and retrieval system, without written permission from the publisher, except for the inclusion of brief quotations in a review.

ISBN: 978-1-942741-54-1

WARNING AND DISCLAIMER

This book provides information about optimizing IT workloads using various technologies and solutions. Every effort has been made to make this book as complete and as accurate as possible, but no warranty or fitness is implied.

The information is provided on an "as is" basis. The author and Hewlett Packard Enterprise Press shall have neither liability nor responsibility to any person or entity with respect to any loss or damages arising from the information contained in this book.

The opinions expressed in this book belong to the author and are not necessarily those of Hewlett Packard Enterprise Development LP.

GOVERNMENT AND EDUCATION SALES

This publisher offers discounts on this book when ordered in quantity for bulk purchases, which may include electronic versions. For more information, please contact U.S. Government and Education Sales 1-855-447-2665 or email sales@hpepressbooks.com.

Feedback Information

At HPE Press, our goal is to create in-depth reference books of the best quality and value. Each book is crafted with care and precision, undergoing rigorous development that involves the expertise of members from the professional technical community.

Readers' feedback is a continuation of the process. If you have any comments regarding how we could improve the quality of this book, or otherwise alter it to better suit your needs, you can contact us through email at hpepress@epac.com. Please make sure to include the book title and ISBN in your message.

We appreciate your feedback.

Publisher: Hewlett Packard Enterprise Press

HPE Press Program Manager: Michael Bishop

CONTENTS

Foreword xiii
Introduction xv

1 **Anatomy of Workloads** 1
   The evolution of workloads 2
      Workload categories 4
      The Internet of Things (IoT) as another workload category 6
      Workloads Framework 7
   Summary 9

2 **High Performance Computing: Business Intelligence** 11
   Solution overview 12
   From business goals to PoC to production 14
      Hardware inventory 16
      Software inventory 19
      Deployment 19
   Summary 20

3 **High Performance Computing: Scientific Research** 21
   Solution overview 22
   How did we arrive at this solution? 23
      Server space 24
      Server power 24
      Server cooling 25
   Hardware inventory 25
   Implementation project plan 28
   Solution validation and success criteria 28
   Summary 28

4 **Scale-Out Workload** 29
   Scaling a single-threaded application 31
   How did we arrive at this solution? 32
   Hardware inventory 33
   Software inventory 34
   Implementation project plan 35

Expertise and skills 36
Solution validation and success criteria 36
    Application architecture 37
    Test application description 37
    Test workload data/results 38
    Analysis and recommendations 39
    Deployment overview 39
    Management tool overview and comparison 40
Summary 41

5 Scalable Storage 43
Case #1: Object storage for unstructured data 43
Why object storage? 44
Solution overview 44
Hardware inventory 46
    Storage server (16 Total) 46
    Connector server (2 Total) 47
    Supervisor server (1 Total) 48
Object storage software inventory 48
Sizing calculations 49
Key features of object storage 49
    Erasure coding 49
    Scale-out 50
System management 51
Deployment overview 51
Case #2—Two-site object storage 52
Active/Passive object storage 52
Solution design 53
Summary 55

6 Archiving Large Files 57
Case #1—Archiving large files to disk 57
Solution overview 58
How did we arrive at this solution? 59
    Server and storage 59
    Hardware RAID settings 61
    Controller selection 61
Hardware inventory 64
Implementation project plan 66

    Solution validation and success criteria 66
        Looking ahead 67
    Case #2—Replacing tape with object storage for file archive 67
    Solution overview 69
    Solution details 70
    Summary 72

7 Hosted Desktop Infrastructure (HDI) 73
    Solution overview 73
    How did we arrive at this solution? 76
    Solution inventory 77
        Client configuration 78
        Cartridge configuration 78
        Citrix software 79
        Moonshot software 79
        Moonshot power supplies 80
        System management 80
    Summary 80

8 Virtual Desktop Infrastructure (VDI) 81
    Solution overview 81
    How did we arrive at this solution? 83
        Processor selection 84
        StoreVirtual VSA and Hypervisor Compute Considerations 85
        StoreVirtual VSA performance and capacity 85
    Solution inventory 86
        Software for VDI environment 88
    Summary 88

9 engineering Virtual Desktop Infrastructure (eVDI) 89
    Why eVDI 89
        Business benefits of eVDI 89
    Solution overview 90
        Datacenter considerations 95
    How did we arrive at this solution? 96
        Processor selection 99
        Disaster Recovery (DR) considerations 99
    Solution inventory 99
        Software components 100
    Summary 100

10 Citrix Implementation on HPE Hyper Converged 250 101
Solution overview 102
   Design for Multiple locations 103
How did we arrive at this solution? 104
   HPE StoreVirtual technology 105
   Citrix XenApp Environment 105
Hardware and software requirements 107
   Hardware requirements 107
   Management software requirements 110
   Microsoft licensing requirements 110
   General requirements 110
Summary 110

11 EMR Software on HPE ConvergedSystem 111
ConvergedSystem 700 111
Solution overview 113
Design considerations 116
   Compute platform design considerations 116
   Storage design considerations 117
Management and monitoring 120
   Implementation and support 121
Combined hardware and software inventory 121
Summary 126

12 Hyper Converged System Running Multiple Workloads 127
Solution overview 127
Hardware inventory 129
Software inventory 132
Implementation project plan 133
Solution validation and success criteria 133
Summary 136

13 SAP HANA 137
SAP HANA deployment for nonproduction environment 137
Sizing and design 138
Solution overview 139
   SAP HANA infrastructure input sizing summary 139
   SAP infrastructure design assumptions 140
   Solution components 140

Hardware inventory 142
Deployment overview 144
Summary 144

14 Microsoft SQL Server Scale-Up Workload 145
Summary of Microsoft workload 145
Solution overview 146
How did we arrive at this solution? 147
Hardware inventory 149
Software inventory 150
Implementation plan 151
Expertise and skills 152
Solution validation and success criteria 152
    Application architecture 153
    Application testing 153
    Analysis 156
Summary 156

15 Backup and Recovery of Virtual Machines 157
Solution overview 157
How did we arrive at this solution? 160
Hardware inventory 161
Software inventory 163
Implementation steps 163
    Fiber channel step-by-step setup instructions 164
    iSCSI step-by-step setup instructions 165
Summary 167

16 Hyper Converged Backup and Recovery 169
Solution overview 169
How did we arrive at this solution? 170
Hardware inventory 171
Software inventory 175
Solution validation and success criteria 176
Implementation, deployment, and testing 176
Test workload results 176
Summary 177

17 Disaster Recovery 179
Current and new configurations 180
The new solution overview 181
How did we arrive at this solution? 183
Hardware and software inventory 186
Solution validation and success criteria 189
Deployment overview 190
Validation 190
Summary 191

18 An Asymmetric Approach to Hadoop 193
Design Factors 193
Solution overview 195
How did we arrive at this solution? 197
Compute platform selection 198
Compute node design 199
Drive selection 199
Network setup 199
Management tool selection 200
Management node design 200
Proactive care advanced selection 200
Hardware inventory 201
Implementation project plan 206
Summary 206

19 Hybrid IT: The "Right Mix" 209
Solution overview 210
Solution components 211
Hyper Converged 380 211
Helion CloudSystem 10 Enterprise 212
Cloud Service Automation 213
Operations Orchestration 214
Helion OpenStack 215
Helion Stackato 216
Solution inventory 217
Services and education 220
Summary 221

20 The Power of Connection: The Internet of Things 223
The opportunity with connected, intelligent things 224
Common IoT architecture elements 226
Device connection and data acquisition (Stage 1) 227
Data aggregation (Stage 2) 228
Edge analysis (Stage 3) 229
Deep integration with IT systems and other data sources (Stage 4) 231
Security considerations (supporting all stages) 231
The potential of a connected future 232

21 Memory-Driven Compute and Composable Infrastructure 235
Memory-driven computing 235
HPE Synergy 239
Management of Synergy 242

Epilogue 243

Index 245

Foreword

The industries and customers we serve are going through a significant digital transformation in both their business models and the markets they participate in. These transformations are bringing new opportunities, and are enabling our customers to drive the right outcomes for their business.

Technology has progressed to a point where you can target your environment to meet specific workload needs - and these workloads vary widely! You may have, for instance, a complex problem that will scale-out to thousands of nodes and requires constant tuning, or at the other end of the spectrum, a remote office that needs minimal resources and there is no one to maintain the system.

In this book, HPE experts cover one of the most important technical topics that exists today: Optimizing your environment for individual workloads. The book also provides a rare opportunity to review several tried and tested deployments that have produced excellent results, all based on a mature set of solutions driven by experts in the field with real examples and designs.

From a broader perspective, this book is not simply about workloads. It also demonstrates the fact that Hewlett Packard Enterprise offers not just one or two proprietary platforms but a whole suite of open and flexible IT solutions with innovations such as HPE Synergy and Aruba ready to bridge the evolution to Composable Infrastructure, multi-cloud environments, and the intelligent edge. This aligns with HPE's vision that, in the future, the world will be hybrid, everything will be software-defined, and the edge will explode. The examples you will find in this book showcase the HPE strategy – to make hybrid IT simple, to be the IT of industrial IoT powering the intelligent edge, and having the services expertise to make it happen.

In the end, it's all about the right mix of technology plus the right customer experience, driving the right outcomes.

Best Regards,

Antonio

Antonio Neri
Hewlett Packard Enterprise
Executive Vice President and General Manager, Enterprise Group

Introduction

Why we wrote this book

Workloads are key to every technology decision. Given today's unprecedented growth in IT consumption models to drive business success, more than ever before, the effective deployment of IT resources is critical. Making the right workload placement decisions is pivotal to delivering workloads-based business services in a way that is efficient, flexible, and scalable. To boil it down to a single question, every business operates a series of workloads and wants to understand, what are the ideal platforms to run those workloads on? We wrote this book to help CTOs, data center administrators, systems architects, IT professionals and other technical decision-makers discover some truly practical answers.

How this book is organized

This book covers a sampling of the most common workloads found in companies looking to evolve their data center and IT strategies. It seeks to present the detailed technical information that IT professionals require to identify the right workload platforms to meet their technical needs and achieve their business goals. Every workload has unique characteristics that require specific design features in order to run in an optimized manner. Workloads vary widely, such as scale-out versus scale-up, and have different characteristics, so understanding the fundamental aspects of the workload and crafting a solution that optimizes the operation of the workload is the key to high performance.

These workloads represent a variety of solutions designed by the authors of their respective chapters. The following is a list of the key topics with a short description of each chapter:

Anatomy of a Workload. Chapter 1 is written by a workload expert who provides a broad-based introduction to workloads from an industry perspective including categories of workloads, the data explosion, workload framework, and other high-level related topics.

Scale-Out. Chapters 2-4 cover this topic from three different angles: High Performance Computing (HPC) for actuarial modeling; HPC for scientific research; and scale out over many servers to improve the IT infrastructure.

Object storage and large file storage. Chapter 5 looks at using object storage software to archive petabytes of data across a distributed system, while Chapter 6 focuses on large files stored on disk using custom backup software. Each of these chapters include two case studies and their design outcomes.

Desktop Solutions: Chapters 7-10 cover four different workloads: Hosted Desktop Infrastructure (HDI) in which HPE Moonshot cartridges are devoted to a desktop; Virtual Desktop Infrastructure

(VDI); engineering Virtual Desktop Infrastructure (eVDI); and one chapter that focuses on Citrix. Many firms that had been testing various desktop solutions are now deploying such environments on a large scale and these four chapters provide a good sampling of solutions.

Electronic Medical Records (EMR): Chapter 11 shows an EMR workload solution using HPE ConvergedSystem 700 which is key to the strategy in many healthcare environments.

Hyper Converged System on Virtual Machines. Chapter 12 shows the design and implementation of an HPE HC 250 system to support a variety of virtualized applications.

Scale-up: Chapter 13 covers a SAP HANA solution on an HPE Superdome X; and Chapter 14 covers Microsoft SQL Server 2016 on Superdome X. Scale-up applications require features such as large memory capability one of the many benefits of Superdome X covered in these chapters.

Backup and Recovery: Chapters 15-17 cover three different backup and recovery-related solutions: Chapter 14 shows a way to deploy backup and recovery of virtual environments with Veeam and the requisite backup hardware; Chapter 15 covers the implementation of a Hyper Converged system with HPE StoreVirtual; while Chapter 16 details a Disaster Recovery (DR) solution that focuses on the storage aspects of DR.

Hadoop: Chapter 18 covers an asymmetric approach whereby compute and storage can be scaled independently of one another.

The last three chapters of the book focus on the workloads of the future. They do not cover a specific workload example but instead are high-level topics that highlight strategic choices for optimizing IT operations, including insights into future technologies that will apply to many workloads.

Hybrid IT: Chapter 19 looks at the background and strategic choices for identifying the "Right Mix" of applications to run in your private cloud and public cloud when creating and migrating to a Hybrid IT environment.

Internet of Things (IoT): Chapter 20 introduces IoT at a high-level to give a window into the business opportunities in this fast-growing technical area.

Composable Infrastructure: Chapter 21 covers advanced technologies such as HPE Synergy that is available today. Synergy provides a fluid pool of IT resources suited for many workloads and should be evaluated for any environment.

What you will learn

As you can see from the list of chapters, there are a wide variety of workloads covered as well as an overview of related topics such as IoT and Hybrid IT. This book will help define requirements and make the right choice of tools and technologies. Since each chapter focuses on a particular workload, it is designed for you to reference your specific areas of interest.

The layout of the chapters is similar in that most cover key aspects of the workload including a solution overview, the components used in the solution, the key considerations used to craft the solution, and so on. Most of the chapters include a Bill of Materials (BOM) so that you can see the physical components included in the design. The BOMs are indicative of the type of workload being featured. A detailed customer needs assessment with a professional IT consultant is recommended to evaluate the exact BOM required in a specific instance.

This is a "blue collar" book in that that chapters don't provide a lot of background on the workload topic but instead delve into the solution quickly. The BOM and rack diagrams give a clear vision of the solution implemented and some of the nuances related to the design. In some chapters, for instance, a wiring diagram is covered because it relates to the specific manner in which connections are employed in order to achieve high availability.

Some of the solutions were crafted over months and many iterations of the solution may have been needed to arrive at finished design. In many cases, this was a long and arduous process, that has been distilled in this book down to some essential ingredients. In the end, however, a solid design was produced and that is what you are seeing summarized in each chapter. We hope that this information will help you select the best platforms and leap the hurdles of designing and implementing powerful and effective workload solutions.

About the Authors

This book consists of submissions from many authors who crafted the solution covered in their respective chapter. I worked closely with the authors to produce a book that comes as close as possible to reading as if one author produced the entire book.

Primary Author

Marty Poniatowski is a Senior Director at Hewlett Packard Enterprise in the Business Development, Enablement, Solutions, and Technology (BEST) organization. He manages the presales technical experts in the U.S. East for many disciplines that represent the portfolio of HPE services and products. His focus is on the strategic initiatives of hybrid IT, intelligent edge, data-driven results, and many related technologies.

During his career Marty has authored 18 books on IT topics. Marty holds graduate degrees in Information Technology (New York University) and Management, and an undergraduate degree in Electrical Engineering.

Guest Authors

Mike Brito is a Cloud and Software-Defined Data Center (SDDC) technical sales team leader at HPE. Mike focuses on leadership, team building, creating business value through technical solutions, helping customers discover new opportunities, improving efficiency, and customer engagement. He is a graduate of Roger Williams University and Columbia University.

Steve Haldeman is currently Director of the HPE Americas Strategic Solutions Architecture team providing expert, technical presales solution services across key industry workload and solution areas including the Internet-of-Things (IoT). As a 35 year veteran with HPE, Steve had held many senior technical contributor roles including those in software development, project management, as Solution Architect for Unix systems, advanced systems consolidation, internet-based commerce and high-availability/mission critical system design and implementation. In addition, he has led teams of Solution Architect teams in many of these areas. Steve is an active HPE representative in the area of IoT to industry groups and with customers

John Tsang leads the Americas workload business strategy and go to market to unify a common cross functional approach to workloads & solutions. Leveraging the collective resources of the presales organization and worldwide and regional business units, John has been executing new PAN HPE sales plays and initiatives around the priority workloads.

Most recently John has taken on expanded leadership role to drive the IoT business for North America sales. His responsibilities include leading the IoT go-to-market activities which includes sales, marketing and operations readiness to build the IoT business. John has held numerous leadership roles in end user and channel sales, marketing and business development and joined HPE through the 3Com acquisition.

John graduated with a Bachelor of Business Administration from Loyola University of Chicago and continues to expand his education in coaching and leadership.

Chapter Authors

The following is a list of authors in alphabetical order:

Gary Allard is a Senior Solution Architect at HPE. Gary specializes in scale-out solutions designed for state government, local government, higher education and healthcare institutions.

Jeff Dabal is a senior Pre-Sales Channel Architect covering the enterprise solutions and services portfolio at HPE. Jeff works closely with HPE partners to provide enablement and related activities.

Michael Fahey is a Solution Architect at HPE focused on commercial and enterprise accounts. He specializes in storage technology and solutions.

Vikram Fernandes is a Senior Solution Architect at HPE focused on strategic customer accounts specializing in application and performance characterization.

Kevin Gildea is a Solution Architect at HPE focused on the Service Provider accounts. He specializes in distributed systems and scale-out design. Kevin contributed to multiple chapters in this book.

Rob Goldstein is an Enterprise Architect at HPE, where he supports some of HPE's largest customers by helping them solve challenging business problems through technology.

Anton Hagens is an Enterprise Architect at HPE focused on Commercial and Strategic Accounts with a focus on high-performance computing that meet the clients business outcomes.

Beth O'Malley is a Solution Architect at HPE. Her focus is Channel partner technical enablement that represents the portfolio of HPE services and products focused related to storage.

Zvadia Hibshoosh is a Solution Architect at HPE focused primarily on Industry Standard Servers (ISS) solutions in the Financial Services Industry (FSI).

George Riemer is an Enterprise Architect at HPE. George represents the full portfolio of products and services working directly with HPE enterprise customers.

Taras Rudko is a Presales Solution Architect in the Northeast for Hewlett Packard Enterprise specializing in storage workloads in virtualized environments.

Mary Alice Sallah is an HPE Solution Architect, working primarily with enterprise accounts in the Northeast to design workload-optimized compute infrastructure. Mary Alice submitted two chapters in this book.

Chuck Strobel is Master Solution Architect for HPE's Mission Critical Servers. Chuck is focused on HPE's scale-up Intel-based Servers, SAP HANA Converged Systems, and Integrity HP-UX Server solutions.

Brenda Suárez-Marill is a Solution Architect at HPE. She currently supports some of HPE's largest customers by helping them solve their most challenging technical problems to satisfy business requirements.

Tom Wilson is a Senior Infrastructure Architect for HPE's Enterprise Solutions and Architecture team. Tom focuses on Client Virtualization solutions including Virtual Desktop Infrastructure (VDI) and Server Based Computing (SBC).

Eric Wise is a HPE Master ASE focused on HPE's scale-up Intel-based Servers, SAP HANA Converged Systems, and Integrity HP-UX server solutions.

Benjamin Wold is a Storage Solution Architect at HPE. He specializes on storage workloads in virtualized environments.

Joseph Yanushpolsky is an Enterprise Solutions Architect with HPE Northeast team specializing in design, sizing, and tuning of complex infrastructure and application environments. Joseph contributed portions of two chapters in this book.

1 Anatomy of Workloads

INTRODUCTION

Establishing and running a top performing business requires not only a reliable but also a dynamic information and operations technology infrastructure to handle the demands of all the functional groups within an organization. The focus and commitment from the C-suite within a business to drive differentiated customer outcomes while doing so with higher operational efficiencies puts considerable demands on the information technology organization. With what we are now calling the Idea Economy, there has never been a time in our history where you can easily turn an idea into a business model and do it faster than ever before. This in turn can also bring more challenges to businesses where traditional competitors may not be the major threat anymore, but rather new upstart organizations that are created overnight and cause tremendous disruption.

This era of digital everything is transforming how information technology is consumed with an even a bigger focus around driving business outcomes. In order to achieve these outcomes, information technology solutions must be mindful of the key transformation areas of cloud computing with hybrid infrastructure, big data, mobility including workplace productivity, and security that have provided a perfect storm for this wholesale digital revolution. From cloud and/or multi-cloud computing to even bigger data with the Internet of Things (IoT), more modern approaches to handle and make meaning of this vast amount of structured and unstructured data is key. With the digital collaboration in the mobile workplace and the expansion of what we call the Intelligent Edge, having the levels of security to protect all the applications being accessed from the edge to data center requires intelligence and dynamic allocation of resources across the infrastructure layers.

At the same time, it is not enough to have a purpose-built appliance anymore with server or storage solutions handling one or two workloads. Composable and converged hardware platforms are now required to handle a diverse set of workloads and also have enough intelligence to recognize which workloads are prioritized over others to maximize efficiencies and productivity. Software-defined architectures enable the handling of the vast amount of workloads that have a tremendous influence on the creation of hybrid IT environments that we are adopting today.

Depending on whether the audience is an engineer, IT professional, or business leader, the terminology and taxonomy of what a workload is can vary. The basic definition of workload in computing terms is the amount of processing that the computer has been given to complete in a certain time and space. The workload consists of some level of application programming running on the computer and usually some number of users connected to and interacting with the computer's applications. According to TechTarget in 2006, a defined workload can be specified as a benchmark when evaluating a computer system in terms of performance (how easily the computer handles the workload), which in turn

CHAPTER 1
Anatomy of Workloads

is generally divided into response time (the time between a user request and a response to the request from the system) and throughput (how much work is accomplished over a period of time).

The evolution of workloads

As computing continues to evolve from centralized mainframe technologies in the 1970s and 1980s to distributed computing with client server technologies in the 1990s; to the 2000s' expansion of the Internet and with mobility currently driving cloud computing in the 2010s, how we move, access, and process workloads become front-and-center. As the next decade approaches, the explosion of IoT with increasing sensor technologies being embedded in everyday objects, the speed of how we handle the massive amount of data generated through IoT to make real-time actionable decisions through various applications and platforms securely will bring together the lines of information technology and operation technology like never before.

At its core, IoT is just another differentiated workload and, from a user perspective, it is all about gaining access immediately to information and applications from any device and being able to do something with it anywhere and at any time. This progression is depicted in Figure 1-1.

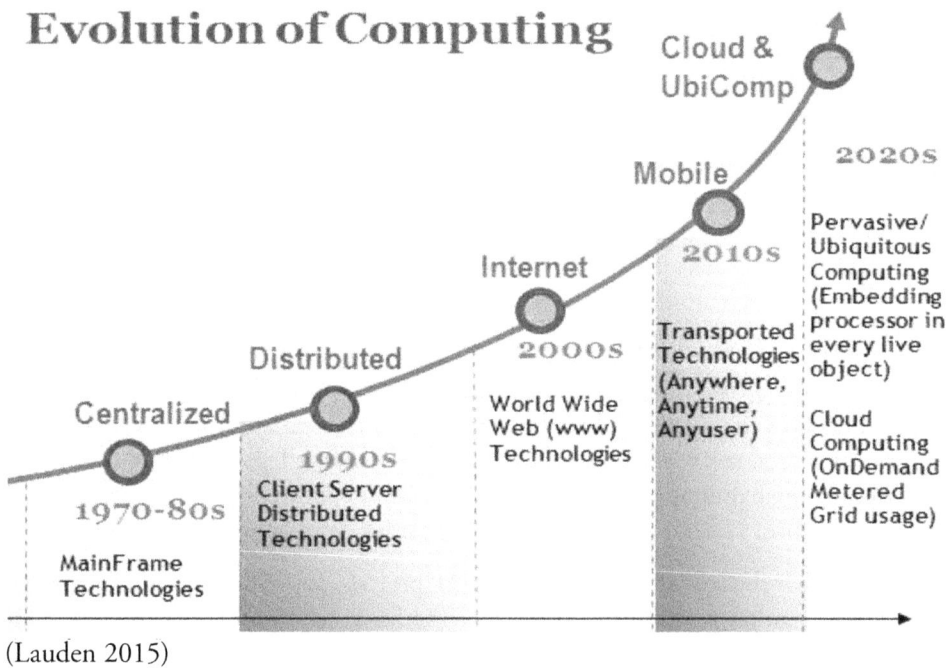

(Lauden 2015)

Figure 1-1 Evolution of computing

There is continued proliferation of mobile devices and growing use of new mobile collaboration tools. This is taking place in industries such as retail, financial services, manufacturing, and will continue to transition certain workloads and applications from on-premises to off-premises computing. The need to address workloads from a hybrid cloud and infrastructure perspective will continue to grow. Enabling on-premises private and hybrid clouds for workloads will be essential as customers prefer to have a hybrid environment to provide more flexibility and security for users. To better understand the shift of workloads to cloud environments, Figure 1-2 defines the various compute and multi-cloud models.

Compute Environment	Definition
On-premises dedicated	The server infrastructure exists on-premises, is dedicated to and optimized for a single application, may or may not be virtualized, but is not cloud-based.
On-premises private cloud	The cloud infrastructure exists on-premises and is provisioned for exclusive use by a single company or institution. It is owned, managed, and operated by the company or institution.
Off-premises hosted or managed private cloud	The cloud infrastructure exists off-premises and is provisioned for exclusive use by a single company or institution. It may be owned, managed, and operated by the company, a third party, or some combination of them.
Public cloud	The cloud infrastructure is provisioned for use by multiple companies or institutions. The cloud infrastructure is owned, managed and operated by the cloud provider at its site.
Hybrid cloud	The cloud infrastructure is a composition of both private and public cloud infrastructures that remain unique entities, but are bound together by technology that enables data and application portability (e.g., cloud bursting for load balancing between clouds).

(Source: IDC, 2016)

Figure 1-2 Cloud models

The challenge for many organizations will be to determine which workloads should run in what environment. Even more importantly determining the workload portability and suitability with the infrastructure options that can enable the various multi-cloud options and different consumption models. Recent research conducted by the TBR group in Figure 1-3 illustrates that more customers tend to perceive on-premises private cloud as more effective to scalability and having more ability to secure and integrate.

CHAPTER 1
Anatomy of Workloads

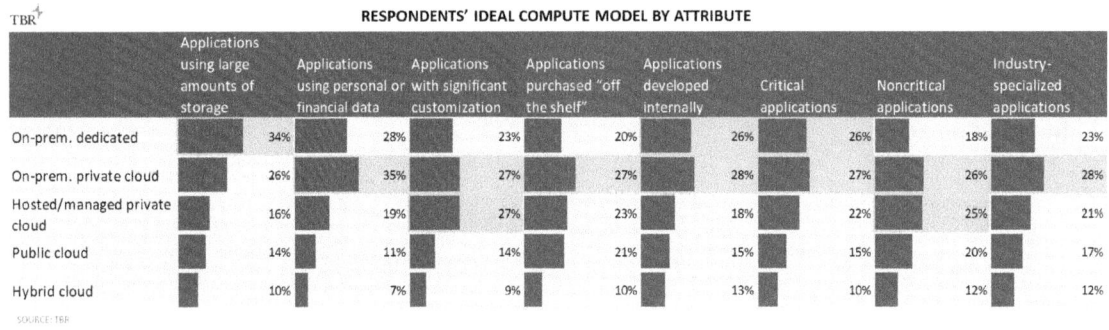

Figure 1-3 Cloud workload comparison

This key understanding reinforces the fact that hardware infrastructure still matters and the multiple ways in which workloads are consumed in this growing hybrid IT environment requires organizations to scale-out and optimize their infrastructure based on business needs. Essentially, workloads will define the solution options for an organization.

Workload categories

With the general context of workloads defined as it relates to compute, the next level of understanding is to define the various workload categories. This categorization can provide more clarity on the simple-to-complex workloads and how it can impact sizing, and choosing, architecting, and implementing the right hardware platform to run for each specific workload. The workload taxonomy mapping (July 2016) from the IDC group offers a detailed categorization of workloads and helps to define the subworkloads associated with the main application areas shown in Figure 1-4.

Workload Categories	Workloads	Subworkloads
Applications	Collaborative	e.g., Web Conferencing, Instant Messaging, Email, Team Collaboration, Enterprise Social Networks, File Sharing
	Content	e.g., Authoring and Publishing, Content Management, Enterprise Portals
	Business Management	e.g., Enterprise Resource Management/Planning (ERM/P), Customer Relationship Management (CRM), Human Capital Management (HCM), Supply Chain Management (SCM), Financial Management
	Engineering/Technical	e.g., Computer Aided Design, Computer Aided Engineering, Computer Aided Manufacturing
	Vertical-specific applications	e.g. Internet of Things*
Data Management	Structured Data Management	e.g., Relational Database Management Systems (RDBMS), Non-relational Database Management Systems (NRDBMS), Database Development and Management, Data Integration and Access
	Structured Data Analytics	e.g., End-User Query, Reporting and Analysis Tools, Predictive Analytics, Offline Analytics, GIS
	Unstructured Content/Data Analytics	e.g., Content Analytics, Discovery, Search, Text Mining, Cognitive Platforms
Application Development	Application Development and Testing	e.g., Cloud native applications, mobile applications, modernization and migration of legacy applications to the cloud
IT Infrastructure	Compute	e.g., Floating Point, Fixed Point, Integer, Memcached, Video Transcoding
	Networking	e.g., Directory, network data/file transfer, communication, and system data/file transfer
	Security	e.g., Identity & Access Management Messaging Security, Network Security, Web Security, Threat & Vulnerability Mgmt.
	Storage	e.g., Replication, Archiving, Software Defined Storage, Back-up, Proxy Caching
	Systems Management	e.g., Event Management, Workload Scheduling and Automation, Performance Management, Change & Configuration Management, Problem Management
	Virtual Desktop Infrastructure	e.g., Desktop operating system hosted within a virtual machine on a centralized server
Web Infrastructure	Media Streaming	e.g., Video and audio multimedia applications for streaming, including an internet component
	Web serving	e.g., Web utilizes HTTP protocols to accept requests from other servers and then search file systems according to the request

Figure 1-4 Workload categories

Understanding this workload taxonomy offers better clarity around the best options for placing workloads in the right place. As workloads continue to be on the move, there is growing delineation due to security risks and proprietary concerns. This results in more simple workloads that require limited scaling such as collaborative workloads (email, instant messaging, SharePoint, and so on) and other general-purpose workloads that are more suited for off-premises and/or the public cloud consumption model. More complex workloads that require extreme scale, high availability, resiliency such as engineering with scientific research, application and testing that are compute intensive, data management and analytics may be better suited to run on a premises in a hybrid cloud implementation.

The Internet of Things (IoT) as another workload category

As the 3Cs of connectivity, compute, and control continue to grow capabilities closer to the edge of the network, more data and the need for device management, integration, and data analytics platforms will create additional types of workloads. IoT furthers the ability to add more intelligence and more connectivity to the objects that surround us. This includes data from the factory, the home, transportation units, vending machines, and virtually any device. These workloads of "Things" as in IoT can be massive containing data levels the likes that have never been seen in history as shown in Figure 1-5.

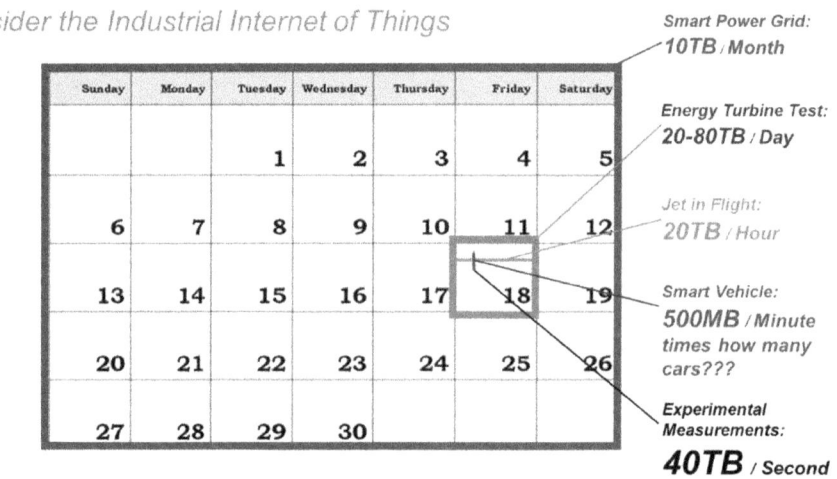

Figure 1-5 IOT Data Size Sampling

As IoT use cases continue to expand, there will be two dynamics at play: Information Technology (IT) and Operational Technology (OT). OT covers the spectrum of systems that deal with the physical transformation of products and services. They are task-specific systems, are highly customized for industries, and considered mission-critical. They typically fall under the domain of engineering. In the world of OT, the end-point being managed is often a physical asset such as pumps, motors, conveyors, valves, forklifts, locomotives, and so on, where these "things" come in all shapes, sizes, level of complexity, versions, and vintage. Most of the software applications in OT's portfolio are "thing-centric" in the sense that they help "make product" by controlling the physical equipment with a great deal of precision and safety, where the human's role is supervisory as automation advances.

Similar to solving a Rubik's cube using many algorithms, the "Age of Algorithmic Businesses" is here in which organizations will be valued not only for their big data but also for the algorithms that turn that data into actions, resulting in improving the customer experience and increase customer impact. The power of the IoT comes from extracting process, business, and customer data that are locked inside enterprises inside devices, inside machines, and inside infrastructure. These data drive visibility into processes, help secure your enterprise against attack, boost the productivity of human and capital assets, and drive profitability by identifying new business opportunities. The following statistics offer justifications as to why IoT workloads will continue to matter for years to come:

- By 2019, 10% of global enterprises will deploy wearable and IoT technologies to monitor the conditions of vulnerable workers in the extended supply chain.
- By 2018, 25% of environmental management decisions will be crowdsourced through IoT sensors.
- By 2020, 38% of installed 6.8 billion IoT devices will be operating with off-grid energy sources.

As a result, the growth of vertical specific applications will evolve to drive specific industry and business outcomes.

Workloads Framework

All the technology in the world is meaningless unless it leads to business outcomes. Organizations need to determine the appropriate solution to run their various workloads and having a common framework that leads to the impact on end users while ensuring the utmost security is paramount. The framework in Figure 1-6 can offer an approach to determining which workloads suit the various platforms and applications based on the business outcomes from each of the key technology transformation areas mentioned earlier.

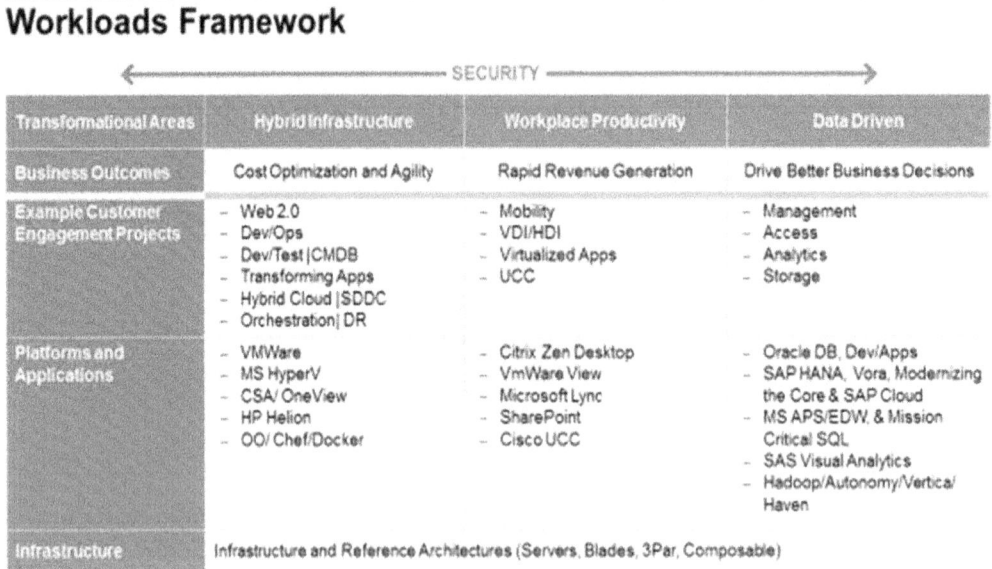

Figure 1-6 Workload Framework

Each of these transformational areas drive a typical major business outcome. For example, data is used to drive better business decisions and as emphasized earlier, data is only as good as how it can be quickly accessed, be able to do effective analysis (especially in terms of IoT where real-time analytics are required to realize the benefits in an operational technology environment), and ultimately, data needs to be managed and stored. There are specific applications and software platforms that enable this access, management, analytics, and storage of data. From there, it is about choosing the best suited hardware infrastructure to enable these applications and workloads securely to drive better business decisions.

A recent study by the TBR group reinforces the applicability of this workload framework by illustrating how enterprises are focused on workloads that are perceived most critical to business outcomes as shown in Figure 1-7.

Ideal Platforms for Optimizing IT Workloads

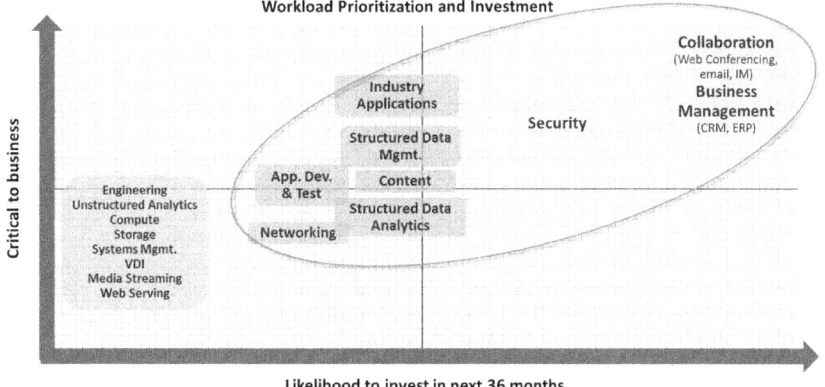

Figure 1-7 Workloads

Summary

As technology continues to evolve and enable more mobility, security, intelligent, smarter everything, workloads will always be on the move and be the main ingredient of how infrastructure solutions will be developed to meet business and customer needs. This solution development requires a common workload framework that organizations can adopt to provide consistency for how IT can bridge technology to solve business problems or even create better efficiencies or expand into new business models and opportunities.

This also applies to the various technology solution providers and their need to be more mindful of approaching their efforts with customers through a workload and business outcomes lens. Not doing so could mean losing their customers' confidence and trust in their ability to really understand their business. Customers recognize quickly when hardware providers position themselves as commodity players versus trusted advisors. Establishing a cross-functional workload framework that addresses a solution go to market that puts the customer use cases first is a timeless approach to being relevant in this idea economy.

References

IDC. (2016). Server and Storage Workloads 2016 MCS. Software Taxonomy Mapping.
Laudon, K. & Laudon, J. (2015). *Management Information Systems:* Managing the digital firm.
Technology Business Research. (2016). TBR-HPE Core Enterprise Vertical Workload Study | July 2016
TechTarget. (2006). Retrieved from http://searchdatacenter.techtarget.com/definition/workload.

2 High Performance Computing: Business Intelligence

INTRODUCTION

High-performance computing (HPC) is the efficient aggregation of parallel computational processing to solve complex problems reliably and quickly. One of its hallmarks is the ability to deliver more processing power in a unit of time, with many servers, than can be delivered by a single computer. Many problems can be solved with HPC including scientific, engineering, and financial modeling. The workload in this chapter is a business intelligence solution for actuarial modeling used for determining product offerings in the insurance industry.

Since actuarial modeling is at the heart of numerous business decisions in the insurance industry, the user community in our solution represented over a half dozen different interest areas that support the various products offered by this financial company. The technical team was given the challenge of meeting or exceeding the following objectives:

- Ensure that the hardware solution would accommodate future growth. The business has already begun to experience slower performance and diminished capacity due to the age of the current infrastructure. More and faster cores were required to replace existing compute platform. The combined business and technical teams wanted to meet the ever-expanding business requirements to be able to model more scenarios and deliver the results of existing and new models more quickly.

- Make certain that power and space were handled as efficiently as possible. This objective made the design concept of shared power supplies and fans across many CPUs very appealing.

- Manage out-of-band access to the servers.

Many platforms were considered, and the HPE Apollo 6000 chassis proved to be an ideal building block to achieve the objectives in this list.

The next section covers an overview of the solution design.

Solution overview

After the lessons from a Proof of Concept (PoC) were considered when crafting the hardware design, the following solution was arrived at for the two locations to support actuarial modeling:

- Location A—Quantity (1448) 3.4 GHz cores spread across 124 HPE ProLiant XL230a Gen9 Servers (compute nodes) located in four racks along with software and services.
- Location B—Quantity (1368) 3.4 GHz cores spread across 114 HPE ProLiant XL230a Gen9 Servers (compute nodes) located in three racks along with software and services.

Each HPE ProLiant XL230a Gen9 Server (compute node) has the following:

- Two Intel® Xeon® E5-2643v3 (3.4GHz/6-core/20MB/135W) Processor Kits
- 96GB of memory
- 2 x 10GbE Ethernet
- HPE Smart Array supporting 4 x 600GB 12G SAS 10k hard disk drives

Each rack contains the following:

- 1–4 HPE Apollo 6000 chassis. Each chassis is capable of supporting up to 10 HPE ProLiant XL230a Gen9 Servers.
- HPE Apollo power shelves to deliver N+N power redundancy to the HPE Apollo 6000 chassis.
- HPE Apollo Platform Manager (formerly known as the HPE Advance Power Manager) to efficiently manage power and compute server systems. The HPE Apollo Platform Manager (APM) provides many capabilities such as automatically discovering hardware, power metering, power capping, and many other features. The APM capability of single Ethernet cable access to all out-of-band management processors (HPE iLOs) on each server with single sign on capability was key to the technical team.
- Two top-of-rack network switches.
- Sufficient in-rack power distribution units (PDUs) to support the in-rack hardware.

Additionally, each site required a head node. After the user submits his actuarial model to the head node, the head node schedules a job on the compute nodes based on availability. The head node is an HPE ProLiant DL380 Gen9 server consisting of the following:

- Two Intel® Xeon® E5-2643v3 (3.4GHz/6-core/20MB/135W) Processor Kits
- 96GB
- 2 x 10GbE Ethernet
- HPE Smart Array supporting 4 x 300GB 12G SAS 10k hard disk drives

Figure 2-1 shows an overview of the design.

HPC Business Intelligence High-Level Design

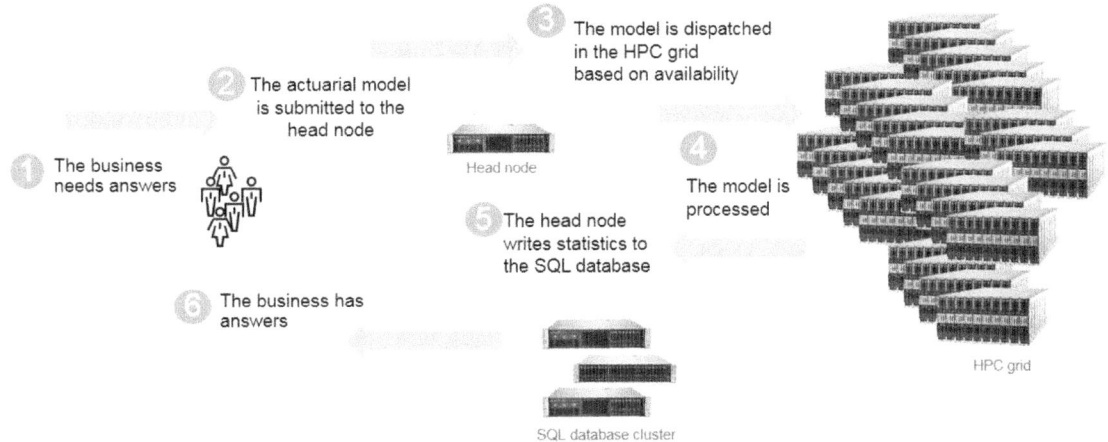

Figure 2-1 HPC High-Level Design

The detailed rack diagram for this design is shown in Figure 2-2.

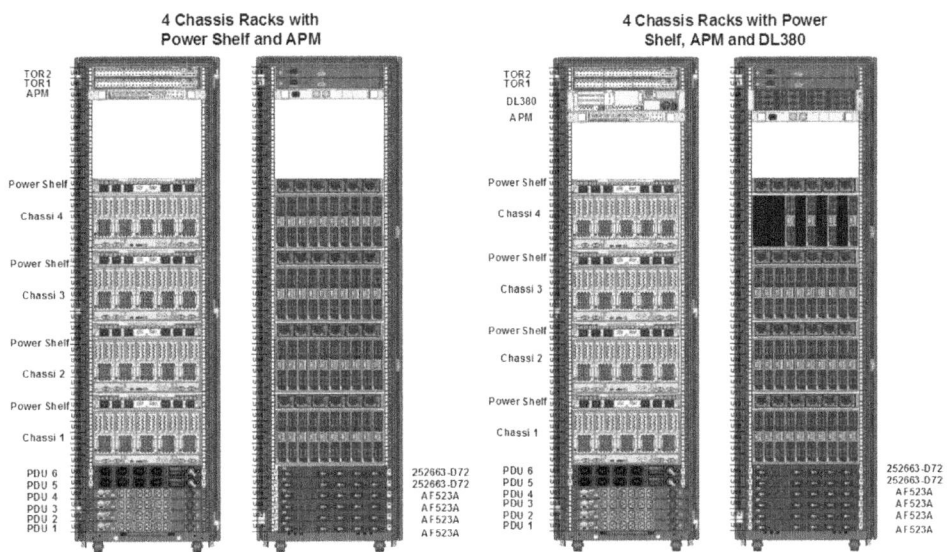

Figure 2-2 Rack Diagram of HPC Solution

From business goals to PoC to production

The hardware had to exceed performance of existing hardware while supporting current and new projects. Existing hardware age ranges from four to seven years. Increased failure rates of this existing older hardware were also a driver to architect and deliver a new solution in a short timeframe. As part of the technical team's due diligence, they investigated deployment of the workload into two different public clouds. One cloud was hosted by the actuarial modeling software supplier and the other cloud was hosted by one of the major players in the cloud market today. Neither deployment was capable of successfully delivering performance requirements. The technical team determined that the hardware had to be on-premises for their highly computational actuarial modeling workload.

Once the decision was made for on-premises hardware, an initial PoC was performed with a single Apollo chassis with four HPE ProLiant XL230a Gen9 Servers. Each HPE ProLiant XL230a Gen9 Servers has the following:

- Two Intel® Xeon® E5-2698v3 (2.3GHz/16-core/40MB/135W)Processor Kits
- 256GB of memory
- 4 x 10GbE Ethernet (configured for 1 GbE for testing purposes)
- HPE Smart Array supporting 2 x 300GB 12G SAS 15k hard disk drives

The entire existing production environment provides 1568 cores, 800 cores in Location A, and 768 cores in Location B. Most production nodes are comprised of the following:

- Two Intel® Xeon® X5670 (2.93GHz/6-core/12MB/95W, DDR3-1333, HT Turbo 2/2/2/2/3/3) CPUs
- 48GB of memory
- 2 x 1GbE Ethernet

For testing, the production hardware was pared down to 128 cores and roughly 1TB RAM spread across 22 compute nodes for the purpose of testing against the new PoC hardware. The initial PoC hardware mimicked the pared down production hardware with 128 cores and 1 TB spread across four compute nodes. The result of the initial hardware was disappointing. The PoC hardware did not outperform the existing production.

While the total number of cores (128) and memory (1 TB) are identical in the two test environments, CPU frequency is different. The Apollo Intel® Xeon® E5-2698v3 (2.3 GHz) deployed in the PoC is 78.5% of the frequency of the Intel® Xeon® X5670 (2.93 GHz) from the existing production environment. While newer CPUs are typically architected to do more in a single CPU cycle, the deficit in the CPU frequency in the PoC hardware proved too much to overcome. Test results showed that most actuarial models have very little improvement in the PoC environment, and some actuarial models actually perform a little worse.

After consulting with the software supplier, the technical team asked HPE to provide additional hardware for the PoC environment. Since the Apollo 6000 chassis on-site had six free slots, HPE decided to provide six new HPE ProLiant XL230a Gen9 Servers that easily plugged into the existing chassis. The technical team in charge of testing requested six servers with:

- Two Intel® Xeon® E5-2643v3 (3.4GHz/6-core/20MB/135W) Processor Kits
- 256GB of memory
- 2 x 10GbE Ethernet
- HPE Smart Array supporting 2 x 300GB 12G SAS 15k hard disk drive

Once the new server nodes were delivered, the technical team was able to execute an unexpected test: upgrade to the environment. The team was pleased with the fast plug-and-play upgrade capability of the HPE Apollo 6000 chassis.

The new PoC environment with only 72 higher frequency 3.4 GHz cores provided desired results. Average runtime decreased substantially for all users. In the end, the fewer higher-frequency cores proved more valuable than a higher number of slower frequency cores deployed in the initial PoC environment.

The technical team still was not satisfied that they had come up with the best solution for the business. The memory in the new PoC exceeded the current recommended 8 GB per core by the software supplier. In order to achieve the most cost-effective configuration, new testing was performed on hardware which reduced the memory to 96 GB per the 12 cores provided by two Intel® Xeon® E5-2643v3 (3.4 GHz/6-core) processor kits. The new configuration of 96 GB per 12 cores followed the recommended 8 GB per core.

A cross-functional set of users performed test calculations and benchmarked against the newly updated PoC environment with 8 GB per 3.4 GHz core and the existing production environment with 2.93 GHz CPUs. Finally, the user community was able to provide results that delighted on a cost-conscious memory configuration. Average runtime decreased substantially for all testers comparing the new to the existing production. A summary of average reductions of various runtimes can be found below:

Table 2-1 Reduction in run times

Comparison of new run times to existing run times	
	Reduction
Application 1	32%
Application 2	40%
Application 3	51%
Application 4	36%

All testers reported successful run execution and significant improvement with the new 3.4 GHz CPUs tested. Besides the faster run times, more of users could be supported on the new PoC environment. Some applications reported that approximately 50% additional users could be supported. These latest successful results differed greatly from the initial PoC hardware with 2.3 GHz CPUs.

The technical team chose to create the new production environment based on the new PoC environment. The number of nodes were almost doubled. Memory was dramatically increased to support the software supplier's current recommendation of 8 GB/core. The new production environment has the ability to support current and future requirements.

Table 2-2 Comparison of new versus existing solutions

Environment comparison							
	Nodes	Cores/Node	Total cores	Speed (GHz)	Total compute (GHz)	Memory (GB)/Node	Total memory (GB)
Existing production*	132	12	1568	2.93	4594.24	48	6336
New production	238	12	2856	3.20	9139.20	96	22,848

* Existing production numbers are approximate due to varying CPU technology deployed over time.

Note:
At the time of testing, the most current CPU and memory technology was deployed, which has since been enhanced.

Hardware inventory

The technical team chose to deploy the hardware in two locations. The distribution of compute allowed for lower wide area network costs by putting compute closer to end-users. The end-users also saw faster response time. Support for the disaster recovery plan was also achieved by deploying in two locations. The architected hardware consists of the following:

- Location A—124 HPE ProLiant XL230a Gen9 Servers located in four racks. Rack 4 contains the HPE ProLiant DL380 Gen9 head node.

Table 2-3 Location A—Nodes per rack

Location A—Nodes per rack			
	a6000 chassis	XL230a compute trays per chassis	Total nodes per rack
Rack 1	4	10	40
Rack 2	4	10	40
Rack 3	4	10	40
Rack 4	1	4	4
Site total	13	n/a	124

- Location B—114 HPE ProLiant XL230a Gen9 Servers located in three racks. Rack 3 contains the HPE ProLiant DL380 Gen9 head node.

Table 2-4 Location B—Nodes per rack

Location B—Nodes per rack			
	a6000 chassis	XL230a compute trays per chassis	Total nodes per rack
Rack 1	4	10	40
Rack 2	4	10	40
Rack 3	1	4	34
	3	10	
Site total	12	n/a	114

Please notice that all racks are not filled to capacity. Expansion in the environment for the unknown requirements of the future was an important goal.

Tables 2-5 to 2-9 detail the building blocks for the Bill of Materials (BOM) for the two locations:

Table 2-5 Rack with Power Distribution Units (PDU) and In-rack Ethernet Cables

Qty	Part Number	Part Name
1	BW908A	HP 42U 600x1200mm Enterprise Shock Rack
1	120672-B21	HP Rack Ballast Kit
1	BW930A	HP Air Flow Optimization Kit
1	BW909A	HP 42U 1200mm Side Panel Kit
1	BW932A	HP 600mm Rack Stabilizer Kit
4	AF523A	HPE Intelligent 17.3kVA/60309/NA/J PDU
2	252663-D72	HPE Basic 4.9kVA/L6-30P/C19/NA/J PDU
9	ZU717A	HPE Ethernet Custom Cable Kit
1	ZU708A	HPE Rack Customization Package

Table 2-5 reflects the rack infrastructure needed to support the maximum configuration of 40 compute nodes and one head node.

Table 2-6 HPE Apollo Power Manager

Qty	Part Number	Part Name
1	741192-B21	HP Advanced Power Manager Kit
4	762048-B21	HP 4M Cnsldtd Mgmt w/Latch Cbl
4	762049-B21	HP 3M McrDB9M Blk Serial Cbl

One HPE APM was deployed per each rack. Please see the **Deployment** section for more information on the HPE APM.

CHAPTER 2
High Performance Computing: Business Intelligence

Table 2-7 HPE Apollo 6000 Power Shelf

Qty	Part Number	Part Name
1	735131-B21	HP Apollo 6000 Power Shelf
1	413379-B21	HP BLc7000 1 PH FIO Power Module Opt
4	757398-B21	HP Apollo 6000 Pwr Shelf 863.3mm Pwr Cbl
6	588603-B21	HP 2400W Plat Ht Plg Pwr Supply Kit
1	765439-B21	HP Apollo 6000 Pwr Shelf Rail Kit

One HPE Apollo 6000 Power Shelf was deployed per HPE Apollo a6000 chassis.

Table 2-8 HPE Apollo a6000 chassis and HPE ProLiant XL230a Gen9 Compute Trays

Qty	Part Number	Part Name
1	736428-B21	HP Apollo a6000 Chassis
10	789917-B21	HP ProLiant XL230a Gen9 Compute Tray
10	799859-L21	HP XL2x0 Gen9 E5-2643v3 FIO Kit
10	799859-B21	HP XL2x0 Gen9 E5-2643v3 Kit
60	810744-B21	HP 16GB 2Rx4 PC4-2133P-R EF Kit
10	700699-B21	HP Ethernet 10Gb 2P 561FLR-T Adptr
10	786087-B21	HP XL2xx Smart Array P440/4G 12W FIO Kit
40	781516-B21	HP 600GB 12G SAS 10K 2.5in SC ENT HDD
10	757401-B21	HP Apollo 6000 Dual FlexibleLOM Riser
10	788126-B21	HP Apollo 6000 PCIe x16 Riser Kit
10	792549-B21	HP XL230a Mini-SAS P440 12G Cbl
10	339778-B21	HP Raid 1 Drive 1 FIO Setting
1	757400-B21	HP Apollo a6000 Chassis Rail Kit
10	512487-B21	HP iLO Adv Track incl 1yr TS&U SW

Up to four HPE Apollo a6000 chassis are deployed per rack. Each HPE Apollo a6000 chassis can support up to 10 HPE ProLiant XL230a Gen9 Compute Trays (compute nodes).

Table 2-9 Head node

Qty	Part Number	Part Name
1	719064-B21	HP DL380 Gen9 8SFF CTO Server
1	719057-L21	HP DL380 Gen9 E5-2643v3 FIO Kit
1	719057-B21	HP DL380 Gen9 E5-2643v3 Kit
6	726719-B21	HP 16GB 2Rx4 PC4-2133P-R Kit
1	724865-B21	HP DL380 Gen9 Universal Media Bay Kit
4	785067-B21	HP 300GB 12G SAS 10K 2.5in SC ENT HDD

Table 2-9 Continued.

Qty	Part Number	Part Name
1	726537-B21	HP 9.5mm SATA DVD-RW Jb Gen9 Kit
1	700699-B21	HP Ethernet 10Gb 2P 561FLR-T Adptr
1	749974-B21	HP Smart Array P440ar/2G FIO Controller
1	786092-B21	HP DL380 Gen9 8SFF H240 Cable Kit
1	733660-B21	HP 2U SFF Easy Install Rail Kit
2	720479-B21	HP 800W FS Plat Ht Plg Pwr Supply Kit
1	733664-B21	HP 2U CMA for Easy Install Rail Kit
1	512487-B21	HP iLO Adv Track incl 1yr TS&U SW
1	H7J32A5	HPE 5Y Foundation Care NBD Service
1	H7J32A5	HPE 5Y Foundation Care NBD Service

One head node was required per location.

Software inventory

Our PoC deployed the following software:

- Actuarial modeling—Milliman® MG-ALFA® v10.x. (Previous version was v8.x)

- Operating System—Microsoft® Windows Server® 2012

- Middleware—Microsoft® HPC Pack 2012 R2, Update 3. HPC Pack 2012 delivers enterprise-class tools, performance, and scalability for a highly productive HPC environment, providing an integrated cluster environment featuring the HPC Job Scheduler, Message Passing Interface v2 (MPI2), and Service-Oriented Architecture (SOA) interactive application support, cluster management, and monitoring tools.

- Database software—Microsoft® SQL Server® 2012 (on head node)

Deployment

Deploying 238 servers can be quite a task. The technical team wanted to avoid being overrun with boxes of technology that might morph from a solution into pieces of an unsolvable puzzle. The team chose to ease the hardware portion of the task by using HPE to rack, stack, and cable the solution. The network team chose to deploy existing network switches. These switches were shipped to HPE for incorporation into the racks so that the cabling could be completed before shipment of the integrated hardware solution. Once on-site, the technical team merely needed to plug in the network and power up the racks.

The technical team also wanted to ease management once the servers were in production. The HPE Apollo Platform Manager (APM) is a rack-level power and system management solution for the HPE Apollo 6000 and other HPE platforms that provides access to an easy-to-use graphical interface and a RESTful API for modern management capabilities at scale.

With APM, you can monitor the health of all servers in the rack while providing an "at a glance" view. HPE APM will automatically discover hardware components and enable bay-level power on and off, server metering, and aggregate dynamic power capping. Dynamic power capping is available at the rack level. HPE APM also provides asset management capabilities for shared infrastructure. You can also link multiple APMs together for a "row view" of APM-based racks. HPE APM features rack-level event logging, RADIUS authentication, an integrated serial concentrator, up to 11 local user accounts, a read only service port and supports SNMP, SSH, Syslogd, and telnet. HPE APM consolidates each nodes iLO. (iLO is HPE's out-of-band management interface for servers.) With a single cable from the data center's management network to the APM, Ethernet access to all server resident iLOs is enabled. A secure single sign on is also provided.

Summary

In the end, the Apollo 6000 solution was able to achieve all the goals of the business and technical teams:

- Provide enough capacity to support a dramatic expansion of current applications and planned new applications anticipated over the next year.
- Easily expand in the future to support unplanned requirements.
- Provide two physical locations for compute processing to support disaster recover capabilities.
- Make the most efficient use of power and space to support the hardware.
- Manage the server node hardware environment at scale.
- Improve hardware support by replacing aging hardware with current hardware that can be supported by HPE with a standard support offering.

3 High Performance Computing: Scientific Research

INTRODUCTION

This chapter covers creating a networked compute infrastructure to run massively parallel scientific programs. Massively parallel compute problems require architectural solutions that allow different parts of a scientific program to be run simultaneously on many individual servers in a coordinated manner. Since these programs often exchange information while running, a high-performance network is also required.

There were many key constraints and considerations when creating the design including the following:

- **High-processing performance**—These scientific problems are compute intensive and usually require servers with many cores and high frequencies. Some compute problems can benefit from additional memory, and some can benefit from Graphics Processing Unit (GPU) acceleration. A balance of servers that have a modest amount of memory, a large amount of memory, and GPU acceleration is needed.

- **High-network performance**—Parallel processing programs exchange messages between servers to communicate and synchronize their work. Since work on an individual compute server could pause waiting for a message, high-network performance is required.

- **Reliable management nodes**—High-performance compute infrastructures require a separate set of nodes that provide workload distribution. These nodes must be reliable and redundant to provide fast and efficient tasks to the high-performance cluster.

- **Cost**—High-performance compute solution sizing is different from standard business solutions in that the cost can be capped by the size of the grant if indeed the installation is in an educational institution that is the case in this example. A design goal is to provide the maximum amount of compute infrastructure that can be purchased under this cap within the bounds of the datacenter's space, power, and cooling.

- **Floor space**—Floor space is limited in this environment, and as such, the solution needs to be as dense as possible.

- **Power and cooling**—Once the density of the solution is increased, the power and cooling requirements also increase. This density needs to be balanced since there are also limits to the amount of power and cooling that are supplied to each rack.

Solution overview

The overall environment of high-performance computing servers and network infrastructure is shown in Figure 3-1.

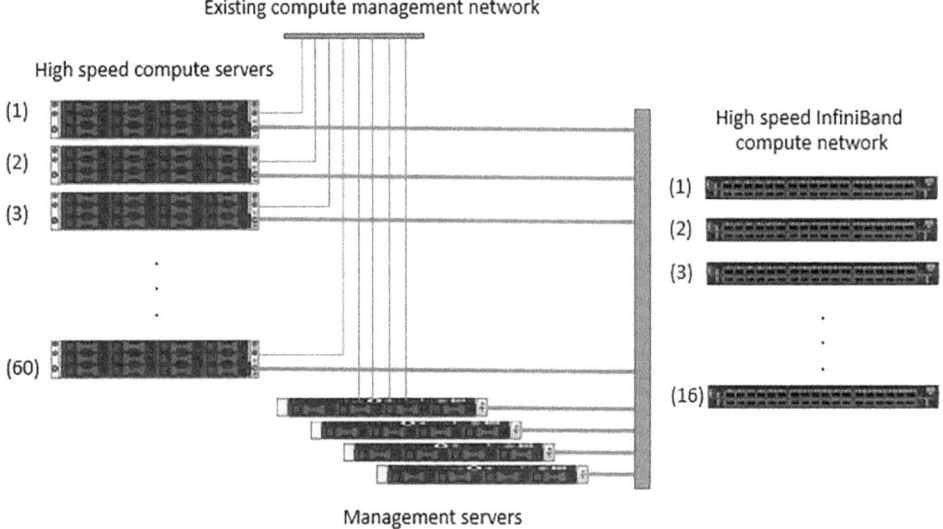

Figure 3-1 High-performance computing design

Figure 3-1 illustrates the solution that consists of 60 HPE ProLiant Apollo 2200 chassis with 230 high-performance computing servers and 4 HPE ProLiant DL360 Gen9 servers acting as management servers connected through 16 Mellanox Enhanced Data Rate (EDR) InfiniBand 36-port switches. This forms a "fat-tree" 56 Gb per second low-latency, high-speed compute network. A network designed as a "fat-tree" logically looks like an inverted tree. The thin network pipes from the individual servers (known as leaf nodes) "tree up" into denser network pipes until it reaches the trunk (also known as the spine) that is a small number of switches that route all traffic. The Apollo chassis complex consists of 192 standard memory nodes, 28 high-memory nodes, and 10 GPU nodes. The following is the list of hardware components:

- 60x Apollo 2200 chassis:
  - 192x Standard memory XL170r servers
    - Dual E5-2650v4 processors
    - 128GB RAM

- InfiniBand EDR port
- 4TB SATA drive
- 1 Gb Ethernet Management port
  - 28x High memory XL170r servers
    - Dual E5-2650v4 processors
    - 512GB RAM
    - InfiniBand EDR port
    - 4TB SATA drive
    - 1 Gb Ethernet Management port
  - 10x GPU XL190r servers
    - Dual E5-2650v4 processors
    - 256GB RAM
    - InfiniBand EDR port
    - 4TB SATA drive
    - 1 Gb Ethernet Management port
    - Two NVidia K80 GPUs
- 4x DL360 Gen9 management servers each with:
  - Dual E5-2650v4 processors
  - 32 GB RAM
  - InfiniBand EDR port
  - 2x 1TB SATA drives
  - 1 Gb Ethernet Management port
  - 10 Gb Ethernet workload port

The next section talks more about the solution.

How did we arrive at this solution?

This section covers three key considerations in creating a design for this solution. The three topics covered in this section are server space, server power, and server cooling.

Server space

The Apollo 2200 solution was considered ideal for the following reasons:

- The Apollo 2200 had an internal capacity to carry either four XL170r dual processor servers with standard or high-memory compute capacity, or two XL190r servers with GPU capacity. This incredible capacity for compute was contained in 2U of rack space.
- Disk requirements were small, as is typical in high-performance compute environment. A single 4 TB disk was configured in every server.
- The Mellanox EDR InfiniBand card provided 56 GB per second of low-latency network capacity.
- The management network was accommodated within each server as part of the solution.

There are many ways to address the space requirements of this project. The following options were also considered:

- *Standard rack mount servers*

   This solution would consume twice the space of the Apollo 2200 solution and was quickly dismissed.

- *Apollo 6000 servers*

   This solution provides similar density, but posed an immediate drawback. The racks provided in the data center were already populated with power distribution units that could not be used by the Apollo 6000 system.

- *Apollo 8000 servers*

   This solution also provides excellent density and superior power utilization; however, the data center could not accommodate liquid cooling.

The Apollo 2200 proved to be the ideal solution.

Server power

A key consideration in this solution was maintaining the power envelope and power configuration that the data center had for maximum power that could be used within a rack.

After running several solutions through the HPE Power Advisor, it was determined that the racks could accommodate no more than 18 Apollo 2200 chassis. As stated above, the Apollo 6000 solution could not use the power distribution units supplied, and the Apollo 8000 could not be used since liquid cooling was not an option.

An example output from the HPE Power Advisor for one rack of this configuration is shown in Figure 3-2:

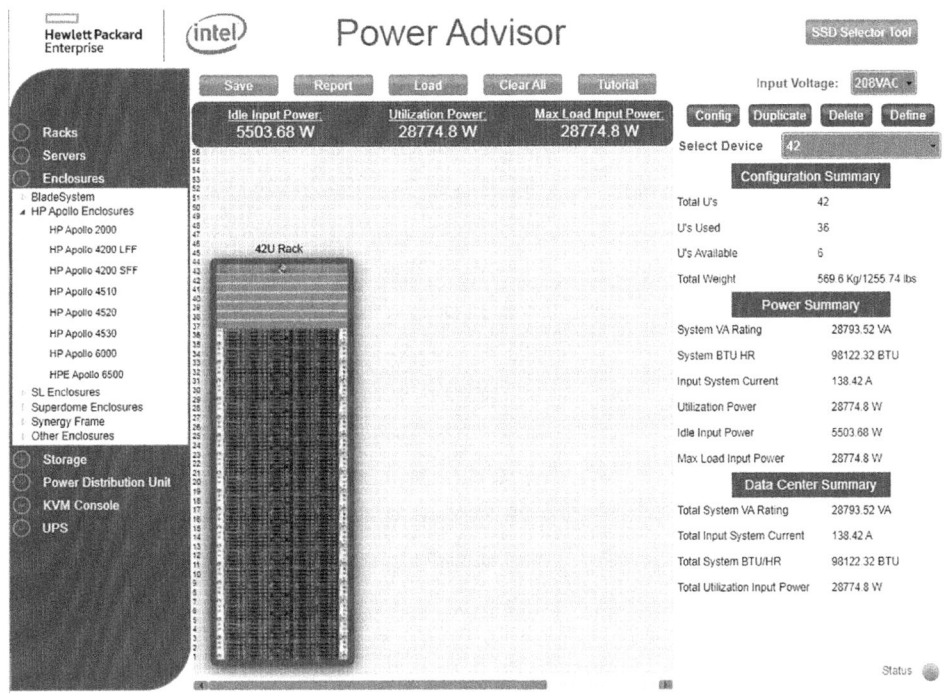

Figure 3-2 Power Screen Shot

Server cooling

Densely designed HPC solutions can generate tremendous amounts of concentrated heat when operated at capacity. As shown in this sample output from the HPE Power Advisor, the single rack portion of the cluster highlighted generates almost 100,000 BTUs/hour. This heat must be dissipated quickly to avoid damaging any equipment within the rack or any equipment located near that rack. The final key consideration in this solution is ensuring that the environmental conditioning is in place to dissipate the heat that is generated by this cluster.

The data center uses forced air cooling. Since the cluster equipment presumes front-of-rack to back-of-rack airflow, venting tiles were placed in front of racks. Air temperature was measured at floor level, and at the top of rack, to assure that cool air was available for all servers within the rack.

Hardware inventory

The following table shows the Bill of Material (BOM) for the individual components of this solution.

CHAPTER 3
High Performance Computing: Scientific Research

Table 3-1 BoM for a single Standard Memory (128GB) Apollo 2200

Qty	Product	Description
1	798152-B21	HP Apollo r2200 12LFF CTO Chassis
4	798155-B21	HP ProLiant XL170r Gen9 CTO Svr
4	850306-L21	HPE XL1x0r Gen9 E5-2650v4 FIO Kit
4	850306-B21	HPE XL1x0r Gen9 E5-2650v4 Kit
32	836220-B21	HPE 16GB 2Rx4 PC4-2400T-R Kit
4	825110-B21	HPE IB EDR/EN 100Gb 1P 840QSFP28 Adptr
4	846783-B21	HPE 4TB 6G SATA 7.2K LFF 512e LP MDL HDD
4	798178-B21	HP XL170r/190r LP PClex16 L Riser Kit
4	798182-B21	HP XL170r Gen9 LP P2 PClex16 R Riser Kit
4	800060-B21	HP XL170r Mini-SAS B140 Cbl Kit
2	720620-B21	HPE 1400W FS Plat Pl Ht Plg PS Kit
1	822731-B21	HP 2U Shelf-Mount Adjustable Rail Kit

Table 3.2 BoM for a single High Memory (256GB) Apollo 2200

Qty	Product	Description
1	798152-B21	HP Apollo r2200 12LFF CTO Chassis
4	798155-B21	HP ProLiant XL170r Gen9 CTO Svr
4	850306-L21	HPE XL1x0r Gen9 E5-2650v4 FIO Kit
4	850306-B21	HPE XL1x0r Gen9 E5-2650v4 Kit
64	805351-B21	HPE 32GB 2Rx4 PC4-2400T-R Kit
4	825110-B21	HPE IB EDR/EN 100Gb 1P 840QSFP28 Adptr
4	846783-B21	HPE 4TB 6G SATA 7.2K LFF 512e LP MDL HDD
4	798178-B21	HP XL170r/190r LP PClex16 L Riser Kit
4	798182-B21	HP XL170r Gen9 LP P2 PClex16 R Riser Kit
4	800060-B21	HP XL170r Mini-SAS B140 Cbl Kit
2	720620-B21	HPE 1400W FS Plat Pl Ht Plg PS Kit
1	822731-B21	HP 2U Shelf-Mount Adjustable Rail Kit

Ideal Platforms for Optimizing IT Workloads

Table 3.3 BoM for a single GPU (256 GB with two NVidia K80 GPUs) Apollo 2200

Qty	Product	Description
1	798152-B21	HP Apollo r2200 12LFF CTO Chassis
2	798156-B21	HP ProLiant XL190r Gen9 CTO Svr
2	850306-L21	HPE XL1x0r Gen9 E5-2650v4 FIO Kit
2	850306-B21	HPE XL1x0r Gen9 E5-2650v4 Kit
16	836220-B21	HPE 16GB 2Rx4 PC4-2400T-R Kit
2	825110-B21	HPE IB EDR/EN 100Gb 1P 840QSFP28 Adptr
2	846783-B21	HPE 4TB 6G SATA 7.2K LFF 512e LP MDL HDD
2	798178-B21	HP XL170r/190r LP PClex16 L Riser Kit
2	827353-B21	HP XL190r Gen9 GD-RT R Riser Kit
2	802855-B21	HP XL190r Gen9 Dul K80/M60 GPU Adptr Kit
2	800061-B21	HP XL190r Mini-SAS B140 Cbl Kit
4	J0G95A	HPE NVIDIA Tesla K80 Dual GPU Module
2	800807-B21	HP XL190r Gen9 Nvidia Enablement Kit
2	838518-B21	HPE Apollo 2000 LFF Bezel Kit
2	720620-B21	HPE 1400W FS Plat Pl Ht Plg PS Kit
1	822731-B21	HP 2U Shelf-Mount Adjustable Rail Kit

Table 3.4 BoM for a single DL380 Gen9 Management Servers

Qty	Product	Description
1	755259-B21	HP DL360 Gen9 4LFF CTO Server
1	818178-L21	HPE DL360 Gen9 E5-2650v4 FIO Kit
1	818178-B21	HPE DL360 Gen9 E5-2650v4 Kit
4	805347-B21	HPE 8GB 1Rx8 PC4-2400T-R Kit
2	652753-B21	HP 1TB 6G SAS 7.2K 3.5in SC MDL HDD
1	766211-B21	HP DL360 Gen9 LFF P440ar/H240ar SAS Cbl
1	789388-B21	HP 1U LFF Gen9 Mod Easy Install Rail Kit
1	749974-B21	HP Smart Array P440ar/2G FIO Controller
1	825110-B21	HPE IB EDR/EN 100Gb 1P 840QSFP28 Adptr
2	720478-B21	HPE 500W FS Plat Ht Plg Pwr Supply Kit

Table 3.5 BoM for Mellanox Switches

Qty	Product	Description
1	834976-B21	Mellanox IB EDR 36P Unmanaged Switch
1	834978-B21	Mellanox IB EDR 36P Managed Switch

The software used in this solution is call custom and related to solving the HPC problem, so there is no software on this BOM.

Implementation project plan

A primary goal for successfully installing a high-performance compute solution into an existing environment is accurately and precisely placing equipment for optimum space utilization, power connection and consumption, cooling benefit, and network cable length and connectivity.

The design team used a floor placement and power model of the rack distribution within the datacenter. Then the distribution of Apollo 2200 and DL360 chassis deployment within the five racks was calculated. This was adjusted to account for the following:

- Mellanox InfiniBand switch distribution
- Maximum power available within each rack
- Available cooling within each rack

At that point, an elevation document was created that shows the correct Ethernet, InfiniBand, and power cable lengths necessary to reach from the servers to the equipment. This paved the way for engineering experts to come on-site to perform installation using the rack distribution diagram and labelling guide for a fast and efficient method for the systems to be installed, cabled, labeled, and powered on.

Solution validation and success criteria

Given the high degree of planning, this solution was installed very quickly. Within days of installation, all servers had been successfully powered on and communication within nodes and within the management infrastructure had been achieved.

As underlying components change, such as processors and memory, and application requirements increase, this solution will require modification.

Summary

This solution is designed to accommodate growth. Based on the high degree of granularity that the Apollo 2200 allows, the growth can occur rapidly or in small increments. This high-performance compute cluster can still grow in compute power, connectivity, and management.

However, these calculations must continually be revisited and revised. Solutions of this type typically have a lifespan of 2–3 years at most before new technology innovations and increased application requirements outstrip the fundamental limitations of current architectures, and the entire solution set must be revisited.

4 Scale-Out Workload

INTRODUCTION

This is a scale-out application that employs density-optimized servers. In this case, like many scale-out workload designs, our goal is to meet the following criteria:

- **Acceptable performance**. The number of threads running per server is a typical goal in scale-out designs and was a firm requirement in this design. Latency, based on the number of threads running per server, was the key measurement of acceptability.

- **Minimizing power and floor space**. Although floor space was not as important in this design, these two factors are always important in scale-out due to the many servers deployed in such applications.

- **Pricing**. There is no getting away from the importance of pricing in a scale-out design due to the number of servers deployed, and this design was no exception. In this case, the existing scale-out environment of older servers will remain in place while the new servers are added to embellish the existing environment and accommodate the anticipated increase in capacity that is required.

- **Deployment within a specified window**. This criteria must be met in order to meet a significant increase in the number of threads running in this environment.

Density optimized Apollo 2000 servers provided the ideal platform to meet these requirements. The next section is an overview of the overall solution including a rack diagram.

SOLUTION OVERVIEW

In order to achieve the objectives listed earlier, an HPE Apollo-based solution with the following components was designed (see Figure 4-1):

- 12 Apollo Servers per rack with the following characteristics:
  - 48 x XL170r servers
  - 196GB RAM

- Dual Intel E5-2697v3 processors
- Dual 10Gb-T
- 120GB SSD for Host OS
- Advanced Power Manager (APM) positioned at the top of rack, providing the following functionality
    - Complete out of band server, chassis, and rack power control
    - ILO port consolidation (2 Ethernet connections vs. 40 connections per rack)
    - On demand real-time power measurement at node, chassis, and rack level in AC and DC
    - PDU-level power outlet control and current measurement
    - Configurable power up dependencies and sequencing
    - Rack Level Static or Dynamic Power Capping
- Top of Rack Switch
- Software stack (see Table 2-3)
- Services overview (refer to project plan later in the document)

These components are shown in the rack diagram in Figure 4-1:

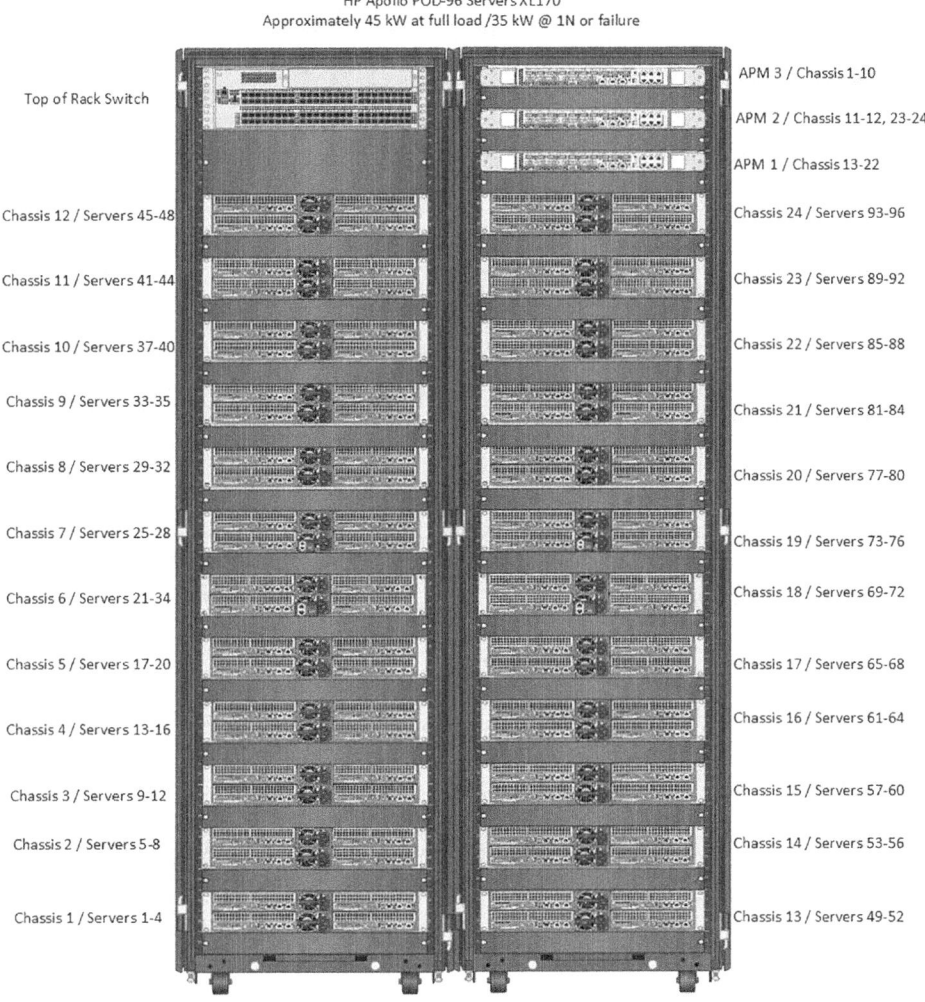

Figure 4-1 Rack diagram depicting the scale-out solution

Scaling a single-threaded application

Each application that is hosted within a scale-out solution has its unique constraints. Historically, mission-critical applications, which were built before the advances in multi-core processors, were designed to maximize use of all available resources during execution. Understanding a profile of these applications is critical to determining the type of multi-core solutions to implement. As an example, an application that suffers from an I/O bottleneck will not benefit from additional processors or memory because the application would be unable to access information in a timely manner.

Application benchmarking is one way to characterize the resource needs of an application. The other would be to include instrumentation into a running application and study the telemetry data that will aid in proposing a tailored solution. The more data that you have, the more deterministic the solution, which implicitly also provides you with the right value at the right cost.

HPE has developed Linux kernel instrumentation tools based on our experience of developing mission-critical operating environments. The deep insight and experience gained in mission-critical application analysis was applied to benchmarking the applications in this chapter. The data sets retrieved over a series of runs using both approaches were instrumental in characterizing these applications and thereby enabled HPE to propose the ideal platform on which to run the applications.

How did we arrive at this solution?

A key method of determining the threads per server in a scale-out solution is to assess the characteristics of the processors. As shown in the table below, we evaluated the following processor options with their respective characteristics:

Table 4-1 Processor Options

Name	Codename	Clock	Cores	Threads	Cache	TDP
Xeon E5-2640 v3	Haswell EP	2.60 GHz	8	16	20 MB	90 W
Xeon E5-2660 v3	Haswell EP	2.60 GHz	10	20	25 MB	105 W
Xeon E5-2690 v3	Haswell EP	2.60 GHz	12	24	30 MB	135 W
Xeon E5-2697 v3	Haswell EP	2.60 GHz	14	28	35 MB	145 W

The number of threads is doubled with hyper-threading enabled so the 2697 v3 supports 28 threads with hyper-threading enabled. The total number of threads supported in our environment looks like this:

28 (threads per processor) x 2 (processors per server) x 1052 (number of servers) = 58,912 threads

Testing was performed to ensure that the number of threads running on each processor met the performance criteria.

 Note
> Since the time that this chapter was written, processor technology has advanced so you will need to evaluate current processor technology as it relates to your scale-out application.

Hardware inventory

The following table shows the Bill of Materials (BOM) for this scale-out application for 1052 servers. As with the processors, and other items previously discussed, technology has advanced since the time of this writing so there are newer server models available.

Table 4-2 BOM for scale-out solution

Item #	Part Number	Product Description
263	798153-B21	HP Apollo r2600 24SFF CTO Chassis
263	798153-B21 ABA	US—English localization
526	800059-B21	HP Apollo 2000 FAN-module Kit
526	800059-B21 0D1	Factory integrated
1052	798155-B21	HP ProLiant XL170r Gen9 CTO Svr
1052	798155-B21 0D1	Factory integrated
1052	793032-L21	HP XL1x0r Gen9 E5-2697v3 FIO Kit
1052	793032-B21	HP XL1x0r Gen9 E5-2697v3 Kit
1052	793032-B21 0D1	Factory integrated
8416	726719-B21	HP 16GB 2Rx4 PC4-2133P-R Kit
8416	726719-B21 0D1	Factory integrated
8416	726718-B21	HP 8GB 1Rx4 PC4-2133P-R Kit
8416	726718-B21 0D1	Factory integrated
1052	665240-B21	HP Ethernet 1Gb 4-port 366FLR Adapter
1052	665240-B21 0D1	Factory integrated
1052	756621-B21	HP 120GB 6G SATA VE 2.5in SC EV G1 SSD
1052	756621-B21 0D1	Factory integrated
1052	798180-B21	HP XL170r FLOM x8 R Riser Kit
1052	798180-B21 0D1	Factory integrated
1052	800069-B21	HP XL170r Rear PCIe Blank FIO Kit
1052	800060-B21	HP XL170r Mini-SAS B140 Cbl Kit
1052	800060-B21 0D1	Factory integrated
1052	HA839A1	HP Fctry Express Svr Sys Custom SVC
1052	HA842A1	HP Custom Image Load Service
1052	HA848A1	HP Firmware Revision Service
1052	HA841A1	HP Custom Asset Tag Service
526	720620-B21	HP 1400W FS Plat Pl Ht Plg Pwr Spply Kit
526	720620-B21 0D1	Factory integrated

Table 4-2 Continued.

Item #	Part Number	Product Description
263	798211-B21	HP Apollo 2000 RCM-module Kit
263	798211-B21 0D1	Factory integrated
1	H7J32A3	HP 3Yr Foundation Care NBD Service
263	H7J32A3 YHE	HP Apollo 2000 Supp
1	HA113A1	HP Installation Service
263	HA113A1 58Y	HP Apollo 2000/4200 Install SVC
1052	BD778AAE	HP iLO Scale-Out incl 1yr TSU Flex E-LTU
1	H7J32A3	HP 3Yr Foundation Care NBD Service
1052	H7J32A3 RYN	HP iLO ScaleOut - 1yr MCC SW Supp
263	611428-B21	HP DL2000 Hardware Rail Kit
1	HA113A1	HP Installation Service
263	HA113A1 5BW	ProLiant Add On Options Installation SVC
22	JG894A	HP FF 5700-48G-4XG-2QSFP+ Switch
44	JG900A	HP A58x0AF 300W AC Power Supply
44	JG900A B2B	JmpCbl-NA/JP/TW
22	UX120E	HP Networks 5810/5800 Installation SVC
22	U4VE4E	HP 3y Nbd Exch HP FF 5700 FC Service
44	JC682A	HP 58x0AF Bck(pwr)-Frt(ports) Fan Tray
22	JG325B	HP X140 40G QSFP+ MPO SR4 XCVR
27	741192-B21	HP Advanced Power Manager Kit
265	762048-B21	HP 4M Cnsldtd Mgmt w/Latch Cbl

Software inventory

There are several modules of software that are essential to the operation of this scale-out solution. In scale-out, the problem is divided among many servers, and software of various types is used to coordinate the work done by many servers to work in conjunction with one another. Table 4-3 shows the software components that are implemented in this scale-out environment:

Table 4-3 Software used to implement this scale-out environment

Item #	Product Description	Additional Details
1	RHEL 6.5	Red Hat Enterprise Linux operating system with custom profiles
2	SICStus Prolog	Commercial prolog runtime with JIT compilation provided by SICStus
3	Custom Application	Scale-out application each instance requiring a corresponding OS thread
4	NFS v4 with RHEL 6.5	NFS v4 to communicate with storage that contains shared files
5	Altiris	Provision Operating system
6	RH Satellite	Host and license maintenance performed using Red Hat Satellite Server

Implementation project plan

Due to the many racks of servers being deployed in a scale-out application, the implementation must be well-planned. In this case, we ensured that as much work was done when the environment was being manufactured as possible. Any hardware work that can be done during manufacturing makes implementation much smoother in scale-out due to the number of servers being implemented. The same is true for software. If it can be configured and tested during the manufacturing process, then this will shorten the field implementation dramatically.

A snippet from a simplified project plan is shown in Figure 4-2. This plan focuses on implementation only and has been reduced to this simple snapshot to make it easier to appreciate the simple fact that scale-out, by definition, requires a lot of coordination due to the sheer number of servers involved.

Figure 4-2 Snippet of Implementation Plan for Scale-Out Solution

Here are some additional highlights of the complete plan:

- The servers were manufactured in Houston and the first block of activities relates to tracking the servers in the factory.
- The racking of servers is being performed on-site at the customer location and the second block of activities relates to racking the servers.
- Installing the servers in the rack requires both cabling and related activities such as Advanced Power Manager (APM) configuration, IP-related activities, and so on.
- The servers were built with a Linux operating system and Red Hat Satellite was used for additional customization.

Expertise and skills

There were several experts involved in crafting this solution including, but not limited to, a team that consisted of people with the following skills:

- Consulting and application analysis: As described earlier the characterization of an application can be done a variety of ways such as running benchmarks, including instrumentation in the application, and so on. The team on this project had tremendous expertise in these areas
- Design validation: The knowledge gained from application characterization had to be translated into a design. A team with knowledge of scale-out and related topics crafted this design
- Implementation team: Both the proof-of-concept environment and, after testing was complete, the production environment were implemented by a team with extensive hands-on experience
- Management tools: A large scale-out environment needs to be maintained, and this many servers cannot be managed individually. Selecting the right tools, testing, and implementing them was a key part of this implementation. Scripting and related expertise is required for the efficient management of such an environment
- Project management: The key to any implementation, not just scale-out, is a well-coordinated plan and attention to detail provided by a project manager. An implementation of this magnitude requires familiarity with all aspects of the application

Solution validation and success criteria

The final design was crafted as a result of on-site testing that was performed to ensure that we met the success criteria.

Application architecture

Scale-out, like other specific workloads, has a unique set of characteristics including those described in this list below:

- CPU intensive
- Memory requirements per instance is known
- Application is sensitive to network latency and its effects can be determined quickly
- Successful transaction time is well known

Although scale-out applications may vary widely, the four bullets in the list are common to all scale-out environments. These bullets are interdependent in that if there is increased network latency, for instance, then the entire scale-out environment performance will suffer. The next sections cover additional topics related to this scale-out implementation.

Test application description

The service application that was benchmarked returns a variable set of results depending on the service initiator. The time taken to complete a successful transaction by the application was known not to exceed more than 1.5 seconds.

Due to the variable result set requested by different initiators for each transaction, the transaction is split into smaller chunks and distributed to a large number of application instances in order to meet the benchmark transaction completion time of less than or equal to 1.5 seconds. The application being CPU-intensive restricts the stacking of application instances to the number of processing threads on each CPU socket.

Network latency is a contributing factor that affects application response time due to data being centralized on a NetApp filer for each application instance. The following rule of thumb was used to calculate the minimum amount of memory that would be required for each server that is part of the operating complex.

Overcommitted:

(2 GB of memory/application instance +30 GB of file system cache + 30 GB for OS-related processes = 192 GB of memory)

Normal operation:

172 GB of memory

The application is a combination of Prolog code calling into C runtime library routines to process rule-based transactions. Each application instance is single threaded relying on an optimized x86 commercial Prolog runtime. As each x86 processor generation has achieved significant performance gains, there is a constant exercise of benchmarking this application across several generations of Xeon E5 processors.

Test workload data/results

A series of benchmark tests were performed each with a differing mix of transaction requests across a set of HPE servers that are part of this cluster.

The tests themselves were carried out in several iterations also known as *baseline tests* and *advanced tests* as defined in the following descriptions:

Baseline tests

As the name suggests, baseline tests run transaction types across the cluster to achieve a baseline result whereby the application response time is in the tolerance limit of less than or equal to 1.5 seconds. The Transactions Per Second (TPS) achieved as part of this baseline was around 9 TPS for each application instance.

Advanced tests

The advanced tests are the number of transactions/sec that are steadily increased from the baseline to a point where response times begin to deteriorate.

Table 4-4 shows the baseline results that the HPE solution had to meet and/or exceed in order to achieve success.

Table 4-4 Baseline for successful results

Baseline Test—Overcommit number of application instances	
Response Time: (ms)	Message Rate: 33 TPS (25K transactions)
Avg Prolog+Handler	1577.668
Avg Handler	18.183
Avg Prolog	1559.484
CPU Utilization:	71%

The response time is measured in milliseconds (ms) and the baseline test achieved a sustained message rate of 33 TPS when a mixture of 25 K transactions were requested by the load generator.

The HPE result set when did not meet the baseline result set during the initial runs as shown in Table 4-5:

Table 4-5 Initial Testing Results

HPE Apollo Test—Overcommit number of application instances	
Response Time: (ms)	Message Rate: 33 TPS (25K transactions)
Avg Prolog+Handler	1891.308
Avg Handler	141.06
Avg Prolog	1750.255
CPU Utilization:	87%

The initial test results did not meet the baseline performance characteristics by 19.88%. After researching the results, we determined that the existing environment used 10 Gb Ethernet network cards and our initial Apollo server configurations had 1 Gb Ethernet network cards in the initial tests. Data for the cluster was centrally stored within storage filers and was accessed by the cluster via NFS shares. The CPUs were at times starved due to the NFS latency attributed to the 1 GbE network connectivity. This event prompted us to upgrade the network interface card to 10 GbE, which complimented the efficiency of our solution and thereby enabled us to meet the performance goal. This highlights the dependency of a scale-out application on any of the performance factors. When the network performance was not met, the ability of the scale-out application to perform well was undermined and could not be made up by a faster processor or other parameter.

Analysis and recommendations

Having met the requirements for baseline test results, we were able then to proceed to the advanced test. This involved pushing the test to the point at which performance began to deteriorate.

The servers in this scale-out solution communicate with storage units that store shared files. We determined that the E5-2697 v3 processor used in this configuration has cache on-chip that provided an opportunity to overcommit the number of application instances on this processor thereby providing an increased cost–benefit realization. The cache minimized the amount of storage unit communication that needed to take place, which allowed this configuration to go far beyond only the baseline performance.

Deployment overview

Initial deployment was challenging due to the large number of servers and the limited time available between time of order and required go-live date. This was due to the specific circumstances related to this deployment but is common when deployments take place. In order to reduce the time spent on each server once on-site in the data center, an HPE service called Factory Express performed many tasks prior to shipping including the following:

1. Basic Input/Output System (BIOS) modifications as defined by this specific implementation
2. Integrated Lights Out (iLO) configuration including initial local user and password, iLO IP address and network configuration, and iLO license
3. Initial Linux operating system installation and network configuration
4. Initial placement of server nodes into Apollo chasses. Each Apollo chassis had multiple server nodes, and the placement of the nodes was defined before the nodes were shipped
5. Placement of asset tags with server serial number provided back to the customer in the form of a Data Entry Identification Document (DEID) report.

Unique values required for each server, such as IP addresses, were provided to Factory Express to be used during the configuration of the servers. Factory Express provided a DEID with data to be used by the customer for additional processing once onsite.

The on-site deployment time is greatly reduced with as much of the configuration and logistic work, such as the asset tag placement, performed prior to shipment. This is important in scale-out due to the large number of servers that are typically deployed.

Management tool overview and comparison

One aspect of this project was determining the right management tools to support the environment. A scale-out environment has many components, which is the very nature of scale-out, so management tools are key to efficiently handle the operational aspects of the environment. This is similar to the previous section in which Factory Express was covered in order to get as much of the work done as possible before equipment arrives on-site.

The following are some key management tools ideal for a scale-out environment:

- **HPE OneView**—an infrastructure automation engine used for many types of hardware that uses a RESTful API.
- **Cluster Management Utility (CMU)**—used in scale-out for provisioning, monitoring, and management of clusters.
- **Systems Insight Manager (SIM)**—manages inventory, health, firmware updates, software updates, and other aspects of a server environment.
- **Insight Remote Support**—remote support tool used to process service events for detected hardware-related activities.
- **Advance Power Manager (APM)**—used to manage power in rack servers that includes such functionality as dynamic rack power capping, Ethernet access to all iLO in the rack, and federation with other APM.

- **Custom Scripts**—because every environment has custom requirements, there is advanced PowerShell capability that can be deployed. There is a RESTful API for iLO cmdlets that are used for iLO management tasks such as updating iLO firmware.

These tools are used in conjunction to achieve specific management goals of an environment. Depending on your specific goals, the right mix of these tools can be selected to result in the most streamlined management environment.

Summary

Building a scale-out workload solution using high-density optimized servers enables you to meet your IT infrastructure needs cost effectively and with only a minimal footprint. Beyond the selection of the right components, both hardware and software, a critical success factor is the expertise of the deployment team and the execution of a well-coordinated implementation plan. When all of these elements come together successfully, as in the HPE Apollo-based solution highlighted here, the transformation to a high performance environment can be achieved with remarkable speed and efficiency.

5 Scalable Storage

INTRODUCTION

Chapter 5 and Chapter 6 together cover the topic of scalable storage using x86 servers. Each of these chapters include two case studies and their design outcomes. The two cases in this chapter focus on using object storage software to archive petabytes of data across a distributed system. The first case study is a single-site object store that uses erasure coding algorithms to minimize the costly storage overhead associated with archiving a petabyte of unstructured data. The second case is a two-site object store that uses an active/passive configuration across two data centers to ensure full system availability in the event of a single data center failure. The first case in this chapter reviews concepts in object storage that appear throughout each chapter.

Case #1: Object storage for unstructured data

This first case covers a distributed system that uses x86 servers for unstructured data storage. Our goal was to build a scalable storage platform for archiving a petabyte of healthcare data. The majority of the unstructured data at this healthcare institution is comprised of high-resolution medical images and videos, genome data, binary large objects, telemetric readings, metadata, patient files, and research documents. This unstructured data often sits on drive or tape for long periods of time without being accessed. A scalable storage solution for this healthcare data was needed to preserve the data over many of years.

To meet the requirements for a large-scale data archive, we designed a distributed storage system using x86 servers with object storage software. This solution is named HPE Scalable Object Storage with Scality® Software. Our goal with this design was to satisfy the following criteria:

- Minimize the total capacity needed for storing unstructured data, in turn minimizing the cost per-terabyte, which is our key cost metric.
- Design a scalable storage system that is distributed across multiple server nodes.
- Deliver a high-degree of data protection without the need for traditional replication and backup processes.
- Provide customizable availability zones and the capability to expand to multiple datacenters.

Why object storage?

The following were the key considerations addressed while developing this storage design:

- Big data challenges—The growth in the volume and variety of unstructured data has outpaced the capabilities of traditional archive methods typically used by healthcare institutions.

- Current methods—Current methods for archiving this type of unstructured data have relied on tape storage or Write Once Read Many (WORM) optical disks that can only be achieved with increasing complexity and risk at petabyte scale.

- Object storage—Storage technologies with intelligence in the software layer have rapidly evolved to address the demands of unstructured data and extreme scale. This intelligence includes new algorithms using erasure coding that can deliver a high level of data protection while simultaneously decreasing total capacity needed.

- Scalable and agile—To expand capacity and performance linearly, software-defined storage is engineered as a distributed file and object platform that can scale out server by server while maintaining I/O operations to the system.

- Implementation—To enable seamless integration of software-defined storage into existing infrastructure and new hybrid cloud environments, a variety of local network protocols and RESTful APIs are supported for different deployment scenarios.

- Transformational outcome—By using software-defined storage healthcare institutions can realize an affordable cost-per-petabyte, scalable performance, and a high degree of data protection. The storage system has the capability to support multiple data types, protocols, operating systems, applications, and datacenter locations.

Solution overview

At the core of this distributed system is a key-value store based on Chord, which is a peer-to-peer routing protocol designed to respond to changes in the systems topology such that a state change in one server is quickly communicated to the other servers in the cluster. This architecture allows I/O operations to be maintained within the system during events such as capacity upgrades, hardware failures, scaling-out servers, software patches, and infrastructure refreshes.

The storage layer of the system provides a set of intelligent services for writing the data across the distributed system, indexing the metadata, monitoring the system's health, and protecting the data. The connector layer of the services provides a file and object interface for applications. The storage layer can be accessed using a variety of data exchange methods, including RESTful APIs, CDMI, SMB, NFS, and CIFS, and others. These software services are all hosted on x86 servers connected through an IP-based network fabric (Figure 5-1).

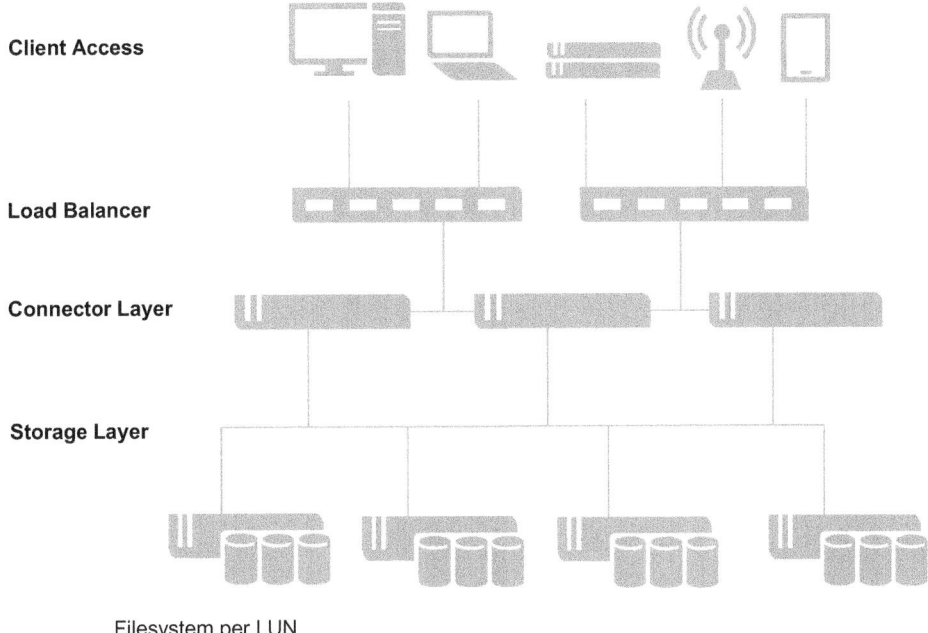

Figure 5-1 Layered design of the Scalable Storage system

The following are the types of servers related to the systems design shown below in Figure 5-2:

- Storage server—Dedicated to I/O operations including write, read, delete, and data preservation tasks, these servers manage the system's interaction with the physical media. Their primary function is to serve requests from the connector servers and locate data using a hash table that is shared across the storage servers. Both HDD and SSD media are needed, with the archive data on HDD and the associated metadata on SSD.

- Connector server—Receives data requests from application servers and coordinates access to the storage layer via a virtual network. The connector servers ingest the data from applications, establish containers, and then transfers the data to the storage servers for final archive.

- Supervisor server—Hosts the management utilities for the Web GUI and CLI. If the supervisor server fails, the system's ability to service requests is not interrupted.

CHAPTER 5
Scalable Storage

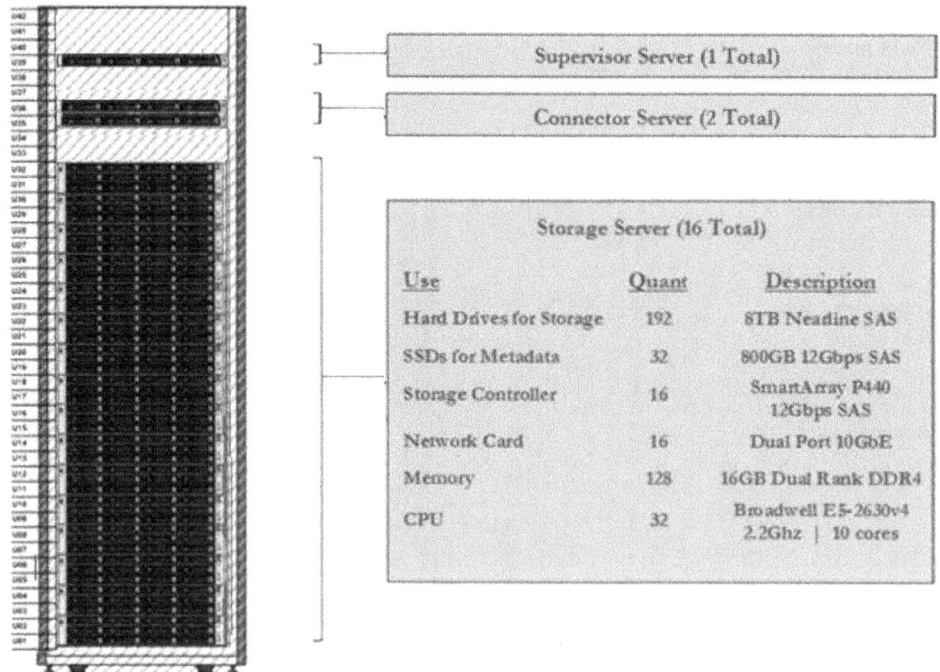

Figure 5-2 Single site 1PB+ usable capacity using x86 servers

Hardware inventory

Tables 5-1 to 5-3 show the Bill of Materials (BOM) for the platform servers. The BOM is divided into storage, connector, and supervisor servers. Both the connector and supervisor servers can be hosted on virtual machines. For processor, memory, and hard drive models in particular, the current technology may have advanced since the time of this writing.

Storage server (16 Total)

For the storage servers, we used density-optimized HPE DL380 line featuring dual 10-core CPUs, 2-port 10 Gb Ethernet and twelve 8 TB drives per server. The HPE DL380 is a compact 2U server with space for 15 LFF drives.[1]

---

[1] The HPE Apollo 4510 server has space for 68 LFF drives per 4U.

Table 5-1 DL380 storage server configuration

Qty	Part number	Description
16	719061-B21	HP DL380 ProLiant Server Configure-to-order Server
16	768856-B21	HP DL380 Gen9 Rear SAS/SATA Kit
16	817933-L21	HP Gen9 Intel® Xeon® E5-2630v4 FIO Processor Kit
16	817933-B21	HP Gen9 Intel® Xeon® E5-2630v4 Processor Kit
128	805349-B21	HPE 16GB 1Rx4 PC4-2400T-R Kit
16	749976-B21	HP H240ar 12Gb 1-port Int FIO Smart Host Bus Adapter
16	665243-B21	HP Ethernet 10Gb 2-port 560FLR-SFP+ Adapter
16	727250-B21	HP 12Gb SAS Expander Card with Cables for DL380 Gen9
16	726821-B21	HP Smart Array P440/4G Controller
16	727258-B21	HP 96w Megacell Batt Cntrl Bd
32	657750-B21	HP 1TB 6G SATA 7.2K rpm Gen9 (3.5-inch) SC Midline HDD
32	762270-B21	HP 800GB 12G SAS LFF 3.5-in SSD
192	793695-B21	HP 8TB 6G SATA 7.2K rpm Gen9 Low Profile Midline HDD
32	720478-B21	HP 500W Flex Slot Platinum Hot Plug Power Supply Kit
16	733662-B21	HP 2U Large Form Factor Easy Install Rail Kit
16	HA454A1-001	Factory Express Package Option for DL380 servers
16	469776-715	HP Add Generic Zpkg Kit

Connector server (2 Total)

For the connector servers, we used HPE DL360 servers with 10 Gb Ethernet adapters. These servers are 1U in height and only require 4 TB total capacity.

Table 5-2 DL360 Connector server configuration

Qty	Part number	Description
2	755259-B21	HP DL360p Gen9 4-Gen9 CTO Server
2	818174-L21	HP DL360 Gen9 Intel® Xeon® E5-2630v4 FIO Processor Kit
2	818174-B21	HP DL360 Gen9 Intel® Xeon® E5-2630v4 Processor Kit
4	665249-B21	HP Ethernet 10Gb 2P 560SFP+ Adptr
2	749976-B21	HP H240ar 12Gb 2-ports Int FIO Smart Host Bus Adapter
2	766211-B21	HP DL360 Gen9 P440ar/H240ar SAS Cbl
4	657750-B21	HP 1TB 6G SATA 7.2K rpm Gen9 (3.5-inch) SC Midline HDD
4	720478-B21	HP 500W Flex Slot Platinum Hot Plug Power Supply Kit
2	789388-B21	HP 1U Gen9 Easy Install Rail Kit

Table 5-2 Continued.

Qty	Part number	Description
2	HA454A1-001	Factory Express Package Option for DL360 servers
2	469776-715	HP Add Generic Zpkg Kit
2	755259-B21	HP DL360p Gen9 4-Gen9 CTO Server
2	818174-L21	HP DL360 Gen9 Intel® Xeon® E5-2630v4 FIO Processor Kit
2	818174-B21	HP DL360 Gen9 Intel® Xeon® E5-2630v4 Processor Kit
4	805349-B21	HP 16GB (1x16GB) Single Rank x4 DDR4-2400
4	665249-B21	HP Ethernet 10Gb 2P 560SFP+ Adptr

Supervisor server (1 Total)

The supervisor server has the same hardware features as the connector servers. While the supervisor can run as a virtual machine on a connector server, we used a separate server to increase the level of fault tolerance.

Table 5-3 DL360 supervisor server configuration

Qty	Part number	Description
1	755259-B21	HP DL360p Gen9 4-Gen9 CTO Server
1	818174-L21	HP DL360 Gen9 Intel® Xeon® E5-2630v4 FIO Processor Kit
2	805349-B21	HP 16GB (1x16GB) Single Rank x4 DDR4-2400 Kit
2	665249-B21	HP Ethernet 10Gb 2P 560SFP+ Adptr
1	749976-B21	HP H240ar 12Gb 2-ports Int FIO Smart Host Bus Adapter
1	766211-B21	HP DL360 Gen9 P440ar/H240ar SAS Cbl
2	657750-B21	HP 1TB 6G SATA 7.2K rpm Gen9 (3.5-inch) Hard Drive
2	720478-B21	HP 500W Flex Slot Platinum Hot Plug Power Supply Kit
1	789388-B21	HP 1U Gen9 Easy Install Rail Kit
1	HA454A1-001	Factory Express Package Option for DL360 servers
1	469776-715	HP Add Generic Zpkg Kit

Object storage software inventory

The software inventory is based on components from Scality® which is a leader in object storage.

Table 5-4 Object storage licenses

Qty	Product number	Description
1	P8Y95AAE	Scality® Installation Package
1017	P8Y89AAE	Scality® Single Site Perpetual License per TB
1017	P8Z01AAE	Scality® 24/7 Software Support License per TB

Sizing calculations

There are many considerations for determining sizing for an object storage archive. Some of the considerations include the average object size, usable capacity needed, and the estimated storage overhead for data protection. For this design, we arrived at the numbers shown in Table 5-5 to meet our requirements for 1 PB of usable capacity for unstructured data with an average object size of 1 MB.

Table 5-5 Requirements for one petabyte of usable capacity

Number of storage servers	16	Servers
Average file/Object size	1	Gigabyte
Total 8 TB hard drives per server	12	Drives
Raw capacity per server	96	Terabytes
Overhead for data protection and file system	33	Terabytes
Usable capacity per server	63	Terabytes
Total 8 TB hard drives	192	Drives
Total raw capacity	1.53	Petabytes
Total overhead for data protection and file system	0.52	Petabytes
Total usable capacity	1.01	Petabytes

Key features of object storage

Object storage has some unique features that make it desirable for storing large files in an efficient and relatively inexpensive way, including the following:

Erasure coding

- To minimize the total raw capacity needed to store the unstructured data, in turn minimizing the cost per terabyte, the distributed system uses erasure coding algorithms across the cluster. Erasure coding performs optimally when objects larger than 10 MB are used. For objects smaller than this size, replication is recommended instead.

- For our design, we chose to implement a combination of 90% erasure coding and 10% data replication. This combination enables the system to maintain the highest degrees of data protection while minimizing storage overhead.

- Had we used 100% data replication, retaining three copies of each object, 3 PB of raw capacity would have been needed to guarantee 1 PB of usable capacity: nearly double the amount required by our design. The number of 8 TB hard drives needed to archive the unstructured data was reduced from 380 drives to 192 drives because of erasure coding algorithms, resulting in 48% less drives needed to store the same 1 PB of data.

- Erasure coding provides an alternative data protection method to replication that is ideal for medium- and large-sized objects. Instead of storing multiple copies of the object, erasure coding stores chunks of objects with parity data. Each object is divided into multiple chunks, x in number, with an additional set of parity chunks, y total. The resulting set of $x + y$ chunks are then distributed across the storage servers. As long as any combination of x chunks is available out of the total $x + y$ chunks, the original object can be completely restored.

- For data protection, the storage servers can be clustered into customizable availability and failure zones. The Scality® software can handle a range of component failures including hard drives, servers, and network connections across datacenter locations. In case of hardware errors, the software isolates faulty components within their respective failure zones.

Scale-out

The Scality® software is deployed on a cluster of storage servers, which can scale to the hundreds as capacity grows over time. The software layer allows for the system to scale-out the access nodes and capacity nodes independently, which allows you to tune IOPS, throughput, and capacity requirements to your specific workload. This is depicted in Figure 5-3.

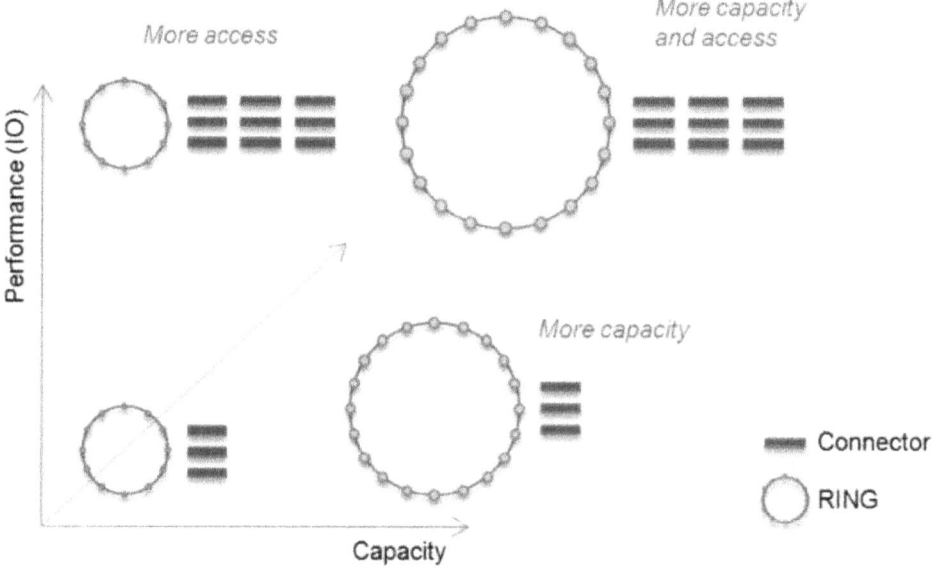

Figure 5-3 Scale performance and capacity independently or linearly

The Scality® software supports file and object system access to the storage layer through the connector servers as previously discussed. For metadata indexing, the software uses a NoSQL database as part of the software layer. This database is shared across the servers and stores the inodes, which is the data structure used to store the metadata for filesystem objects, including time-stamps and data pointers.

System management

To manage the storage infrastructure, each server has an embedded iLO chip that provides out-of-band remote access. The iLO chip collects low-level systems data, monitors the thermal activity of the server, and can be configured to send notifications in the event of hardware failures. The software layer is managed using the supervisor server, which offers a web-based GUI, a scriptable command line interface and MIB traps that can be used with SNMP monitoring consoles. The GUI provides monitoring and management tools to view the entire software stack including logs on hard drives, network connections, servers, and availability zones. Performance statistics and resource utilization can also be retrieved using command line.

Deployment overview

There are some key considerations related to deploying object storage in the following list:

BIOS and Firmware

- Default BIOS settings are recommended with the latest available firmware installed.

RAID Setup

- Since data protection is handled by the software layer, the drives should be configured with RAID 0, Cached I/O, Write Back Cache, and ReadAhead set.

OS installation and configuration

- Install CentOS or RHEL® using the provided KickStart file.
- Setup passwordless SSH between all nodes in the cluster for root.
- Verify that the latest hpsa driver and hpssacli version is installed on all servers.
- Setup and synchronize NTP on each server.
- Verify that all data drives are visible as unpartitioned HDD when "parted–I" is executed.

Networking

- Parameters such as IP addresses, Netmasks, NTP, DNS, and VLANs can be configured on a per-port basis for each server prior to arriving on-site. Network connections in the factory use single 10 GbE links and will not be bonded, although you can certainly bond the links afterward on-site. To connect the servers, you will need a 48-port GigE switch along with corresponding 72-inch cables to reach the top of the rack switch.

Software Installation

- Copy repos.tar from Scality to a directory on all servers.
- Execute "tar xvf repos.tar" on all severs.
- Execute "./install_repos.sh salt-master" on the supervisor server.
- Execute "./install_repos.sh" on all storage and connector servers.
- Run scality_validation.sh from the Scality Supervisor Server to check for errors.

Case #2—Two-site object storage

This case covers an object store that uses an active/passive configuration across two datacenters to support a highly available online archive for retail data. The existing digital content at this retailer is fragmented and spread across a number of disparate systems. This system sprawl has resulted in unpredictable performance and slow queries that often return erroneous data. The goal of this project was to consolidate these disparate systems into an online archive that will serve as a single point of truth for the retailer across their datacenters. The requirements for this two-site online object archive include the following:

- Allow online access to over a petabyte (1 PB) of historic retail data.
- Provide for linear growth up to 4 PB with no loss in performance or availability.
- Improve usability, speed, and quality of query returns.
- Simplify maintenance.
- Provide a unified, enterprise-wide, distributed system that scales-out server by server.
- The solution must be cost-effective, with target of below $1 per gigabyte (GB) of storage. This should include all hardware, software, and 3-year support.

To meet these requirements, we use a distributed storage system using x86 servers with Scality® object software. This distributed system is mirrored from the primary datacenter to another site where the data is backed up and accessible should the primary site go offline. The hardware infrastructure at each location is identical.

Active/Passive object storage

This object storage design features an active/passive configuration between the two datacenters. The data at the primary site is replicated to the secondary site, ensuring high availability in the event of a datacenter failure. Using one stretched distributed system was considered but not accepted due to the

1000+ miles between the two datacenters. Instead, we mirror the distributed system to the second site because this design provides greater fault tolerance and a higher degree of concurrency for online queries.

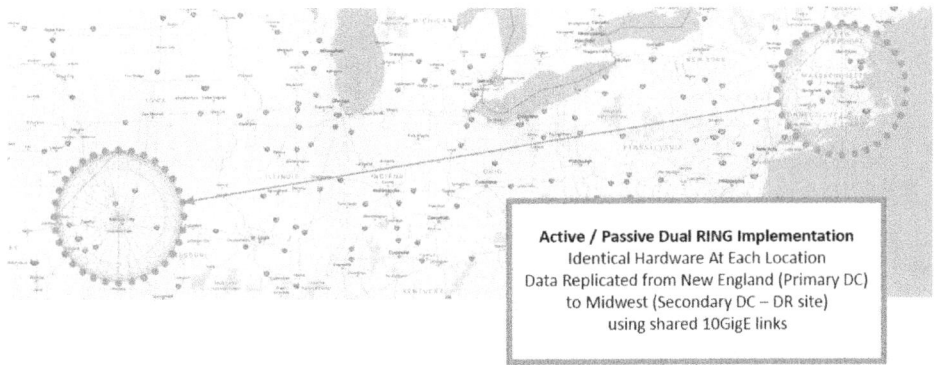

Figure 5-4 Active/Passive object storage mirrored across data centers

Solution design

The hardware infrastructure at each datacenter is identical, and the same object storage software is used at both sites. The following object storage sizer was used to develop a configuration as well as model the future growth of the environment. The sizer takes into account two-site configuration, desired data protection levels, performance expectations, and support requirements. The configurations for 1 PB and 4 PB usable capacities are shown in Table 5-6.

Table 5-6 Two-site object storage for 1 PB and 4 PB usable capacity

	Description	Total quantity 1 PB usable	Total quantity 4 PB usable
Storage servers	Apollo 4510	12	28
CPU	E5-2630 v3 (2.4GHz, 8 Core)	24	56
Memory	DDR4 32GB	96	224
Storage controller	840ar (2 GB Cache)		
NIC	Dual-port 10G	12	28
OS drives	1TB 2.5-inch SATA	24	56
SSD layer	800GB 3.5-inch SATA	24	56
Data layer	8TB 3.5-inch SATA	432	1848

CHAPTER 5
Scalable Storage

Table 5-6 Continued.

	Description	Total quantity 1 PB usable	Total quantity 4 PB usable
Connector servers	DL360	2	12
Software	NFS Connector	2	12
CPU	E5-2630 v3 (2.4 GHz, 8 Core)	4	24
Memory	DDR4 32GB	2	12
Storage controller	H240ar (PCI 3.0)	2	12
NIC	Dual-port 10G	2	12
OS drives	1TB 2.5" SATA	4	24

As covered in detail in the first case study, Scality® software uses x86 servers for the storage and connector layers of the distributed system. In this case study, the HPE Apollo 4510 is used as the storage server of choice instead of the HPE DL380, which was used in the first case study. The Apollo 4510 can fit 68 LFF drives in a 4U chassis. In both object storage case studies, HPE DL360 servers are used for the connector layer.

This systems design meets the requirements for an online archive in the following ways:

- **Scalable**—For 1 PB usable capacity in an active/passive configuration between datacenters, six Apollo 4510 storage servers are needed at each site. Each Apollo 4510 server is filled with 36x 8TB SATA HDD drives and 2x 800GB SAS SSDs. There is room to add up to 30 LFF drives in each storage server, which in total allows for 800 TB usable capacity growth across the system by adding additional hard drives within the chassis. After all of the Apollo 4510 servers are fully populated with 68 LFF drives, the storage system can scale out by adding more server nodes. For 4 PB usable storage capacity, 14 fully populated Apollo 4510 servers are required at each site.

- **High Availability**—Data availability is implemented in the software layer through a combination of replication and erasure coding algorithms. For this archive, the average file size is 500 kilobytes (KB) with 90% of all files being larger than 60 KB. Files 60 KB and smaller are protected by triple replication, whereas larger files are protected by erasure coding ARC (8+4). This means that if one data center is lost entirely and up to three nodes in the other data center are off-line, all the archived data is still available.

- **Performance**—Data access and query performance are stabilized by standardizing on NFS/CIFS access protocols and by using multiple DL360 servers for the connector layer. Performance expectations are met cluster-wide with parallel access to the entire object file namespace. Data is spread evenly across available drive space as more storage servers are added. I/O performance can be increased by adding more connector servers while keeping capacity constant. The separation of the storage and connector layers allows for capacity and throughput to expand linearly or independently. The system can grow into tens of petabytes as needed with no change in performance.

Summary

The two case studies covered in this chapter use x86 servers with Scality® Software to archive petabytes of data across a distributed system. The first case was a single-site object store that uses erasure coding algorithms to minimize the costly storage overhead associated with archiving a petabyte of unstructured data. The second case covered an object store that uses an active/passive configuration across two datacenters to support a highly available online archive for retail data. Although a few design considerations and system choices were different, the same essential hardware infrastructure and object storage software were combined to provide a cost-effective solution that supported the key requirements for scalability, high availability and performance.

6 Archiving Large Files

INTRODUCTION

This chapter covers two similar environments in which local files are used for analytics processing.

In the first case, a disk-based backup solution is used with customer backup software. This allows the local files to be read from the disk-based backup environment for analytics processing.

The second case uses object storage to replace an existing tape backup solution to achieve much faster access of the data for processing. The object storage environment is used with a backup solution but, rather than slow tape technology, a much faster object storage environment is used.

Case #1 — Archiving large files to disk

This first case covers archiving large files to a disk-based storage environment. The large files are video files that are carefully analyzed, and when the analysis is complete, the files can be archived for possible use at a later date. There were many key considerations when crafting the design including the following:

- **Floor space**—Floor space is at a premium in this environment so one of the key goals was to limit the archive solution to a minimum amount of rack space. This solution consumes only 14U of rack space.
- **High capacity**—The video files need to be archived for an extended period of time which resulted in a design of roughly 1 PB of useable storage. This environment can be easily expanded as well by adding storage archive units.
- **High availability (HA)**—HA was in important design factor in this solution. The components of the archive environment must be reliable, and replication is required for a second location to ensure that the archived data resides in two locations. This is depicted in the upcoming block diagram.
- **Performance**—Although archived data may not be accessed for an extended period of time, the video files must be locally available when required. The time to view the video file is not critical and the other design factors take precedence over performance.
- **Cost**—Archived solutions are different from high-performance backup solutions in that the cost is more modest. In addition, this archive environment can be expanded in a modular fashion so that the initial cost is minimized and expansion can take place on an as-needed basis to increase capacity.

CHAPTER 6
Archiving Large Files

Solution overview

The overall environment of video analytics servers connected to the archive solution is shown in Figure 6-1.

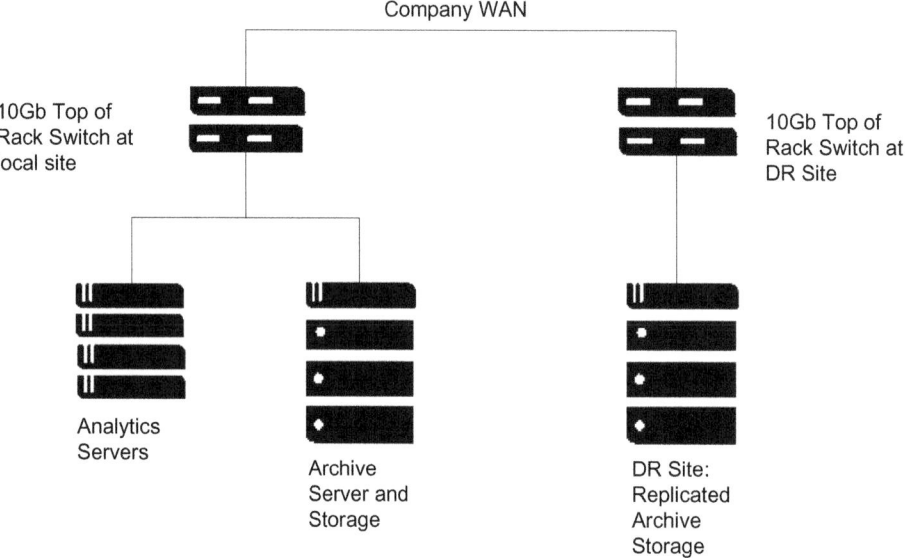

Figure 6-1 High-level video overview

As you can see in Figure 6-1, the solution consists of a ProLiant Apollo 4510 server connected to two D6020 storage enclosures. The Apollo server is used to collect the data from the video analysis servers and then distribute the data to the two D6020 disk enclosures. The following is the list of servers:

- 1x Apollo 4510 server with:
  - Dual E5-2680v3 processors
  - 256GB RAM
  - Dual 10Gb SFP+ ports
  - Dual P440 internal controllers
  - Dual P841 external controllers*
  - 68x 8TB 12G SAS 7.2k drives
  - Separate controller and drives for boot
  - Operating System: Windows 2012

Insight Control licensing with iLO Advanced capabilities—required for virtual media use and remote serial console for troubleshooting

 **Note**
*A special battery controller is needed in order to have four controllers. Customers/partners should check with their HPE team before building solutions.

- 2x D6020 enclosures each with:
  - Redundant I/O modules and cabling per module
  - 70x 8TB 12G SAS 7.2k drives
- Custom backup software designed specifically for this archive solution

Note that the backup software used in this solution is a custom backup software. This archive solution would work with many commercially available backup software solutions.

Figure 6-2 shows the front and back of the rack diagram for this archive solution.

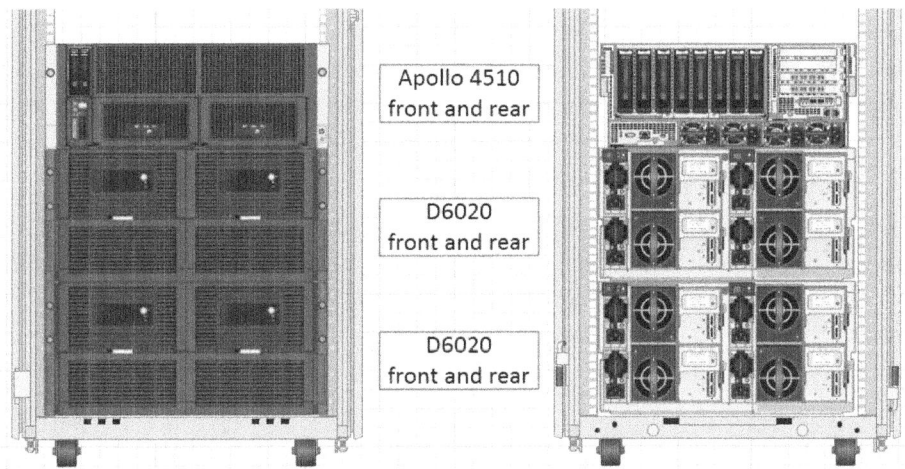

Figure 6-2 Archive solution consuming 14U

How did we arrive at this solution?

This section covers three key considerations in arriving at the ideal solution. The three topics covered in this section are Server and storage, Hardware RAID settings, and Controllers.

Server and storage

Based on the considerations in the Introduction part of this chapter, this Apollo and D6020 solution were considered ideal for the following reasons:

1. Apollo had the highest internal capacity of a ProLiant Gen 9 server, holding 68x 8 TB drives and two boot drives in just 4U.
2. Despite its unusually high storage capacity, it could also include the processor and memory power required for this build, considering the compute needs were limited.
3. The D6020 had unmatched storage density in HPE's storage portfolio, holding up to 70–8 TB drives in just 5U.
4. The D6020 supports daisy-chaining additional enclosures for increased capacity connected to a single server.
5. The D6020 has dual I/O modules, adding high availability in the case of an I/O module failure or port failure on the storage adapter in the server.

The solution focuses only on the archive storage portion of the project. Although the associated servers may have high compute requirements, the server for the archive portion is used simply to access the archive files. This is critical in determining the proper compute-to-storage ratio for the solution.

There are a many of ways to address the storage requirements of this project. The following options were also considered for the video archive:

- Object storage—This is an excellent fit for many archive storage projects as discussed in other chapters. Through erasure coding and replication, object storage provides high availability at the petabyte-scale and maintains a low overall cost because this software solution is deployed on industry standard servers. There were floor space limitations in the project that made the dense storage servers more desirable than object storage.

- 2x DL380 Gen 9 and 2x D6020—A solution utilizing 2x DL380 Gen 9 servers and 2x D6020 enclosures was also evaluated. This provides greater compute density because the compute and storage could scale linearly in building blocks of a DL380 and a D6020. However, as mentioned in this case, the compute needs are not as critical as storage, so this was not necessary. Additionally, a single DL380 Gen 9 and D6020 offers only 656 TB, as compared with the 1.2 PB offered by a single Apollo 4510 and D6020. Without the need for additional compute capability, the Apollo solution also provides less systems to manage compared with the DL380.

- Apollo 4530 and 2x D6020—Similarly, the Apollo 4530 could be evaluated in the place of the Apollo 4510. A single Apollo 4530 offers 3x dual socket compute nodes, allowing for both compute and storage density in just 4U. However, with only 15x LFF drives per node, this solution does not reach the storage density we set forth. At maximum capacity, the Apollo 4530 and D6020 could only offer 920 TB raw storage. This is insufficient compared with the 1.2 PB raw storage provided per Apollo 4510 and D6020.

Working through these alternatives allowed us to identify the ideal solution for this specific application.

Hardware RAID settings

A key consideration in this solution was the hardware RAID settings.

The solution uses an Apollo 4510 and dual D6020 enclosures with the following characteristics:

- 208 Hard Disk Drives (HDD)
- The drives are serial Attached SCSI (SAS)
- Each drive is 8 TB in capacity
- 12 Gb/sec transfer rate
- The drives are 7.2 K

There is no software High-Availability (HA) so Redundant Array of Independent Disks (RAID) is the key to HA in this solution. Performance of the storage subsystem is also a consideration, especially with respect to rebuild times, so we broke the large storage pool into several logical drives, with dedicated spares in each enclosure. We also need to maintain the minimum amount of usable storage. After evaluating the RAID options, we determined that RAID 6 would be ideal for this solution. RAID 6 is known as "double parity RAID," whereby two parity stripes are used on each disk. RAID 6 allows for two disk failures in a group before any data is lost.

The final design has RAID 6 groups with eleven drives each, using nine drives for data and two drives for parity. I will refer to these as 9+2 RAID 6 groups for the remainder of this chapter. As a side note, it is not recommended to have more than eleven drives per RAID group.

This results in five-RAID 6 groups in the Apollo 4510 with two drives available for spares. There are two independent drawers in the D6020, so each drawer will contain three 9+2 RAID 6 groups, with two spares per drawer.

Controller selection

Controllers were critical in the selection process because they perform two key performance functions to:

1. Offload the processing of I/O internal to the Apollo with the dual P440 controllers, and
2. Offload the processing of I/O to the D6020 with the P841 controllers.

Another critical step was selecting the optimal method of connecting the servers to the drives, both internal and external.

When originally configured, the solution had a single P440 controller for the internal drives. This provided only eight paths to share over two internal SAS expanders, with each expander connected to 34 drives.

CHAPTER 6
Archiving Large Files

We adjusted this to have 2x P440 controllers, each with eight paths. Each P440 controller now connects to an independent SAS expander, each with 34 of the drives assigned to it (controller one to drives 1–30, 65–68 and controller two to drives 31–64). The two controllers provide two RAID engines for generating the dual parity drives needed for RAID 6. It also helps to decrease the maximum rebuild time. This arrangement increases performance because the IO traffic is split across two controllers.

For HA purposes, our solution has a controller in the Apollo server dedicated to each D6020 enclosure. The controller is not shared among the two D6020 enclosures. The controller used in this is a P841. This is depicted in Figure 6-3 in which the Apollo sever is shown in the middle of the diagram with a dedicated controller to the D6020 on the top of the diagram and another controller dedicated to the D6020 on the bottom of the diagram.

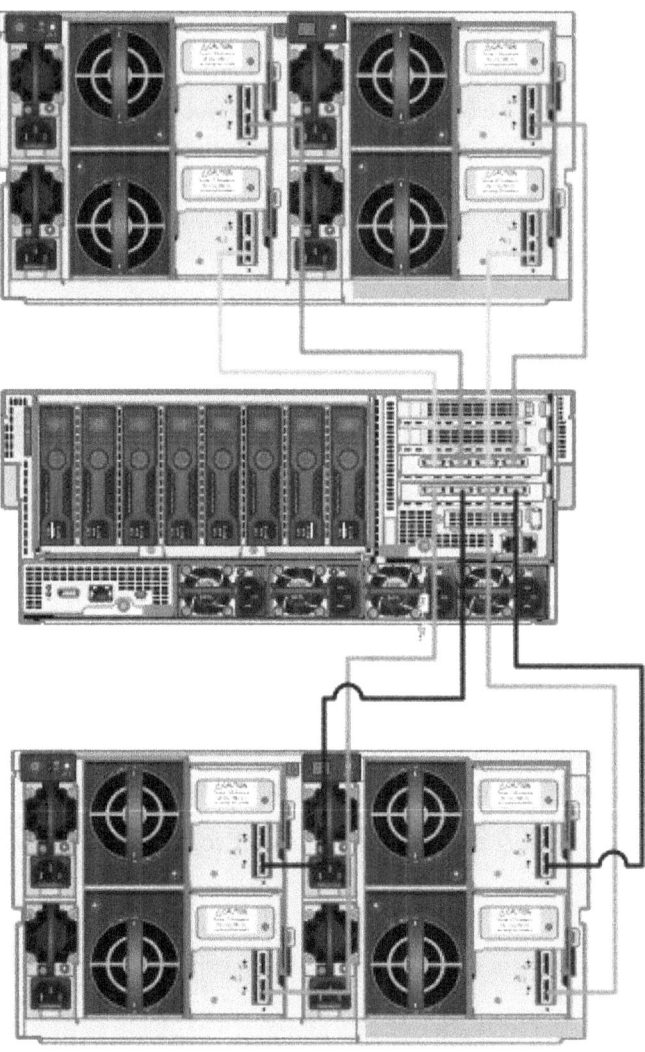

Figure 6-3 Original deployment: Apollo 4510 connected to 2x D6020 enclosures

Note that in Figure 6-3, each P841 controller will be connected via four cables (PN: 716197-B21) to each of the four IO modules on the D6020. This provides redundancy within the D6020: If the first IO module in a drawer were to fail, the second IO module would takeover. In this deployment, the server will have 2x P841s, thus 2x D6020s connected identically per server.

Expansion of this environment will take place so that the initial configuration was crafted with additional D6020 being added in the future as shown in Figure 6-4.

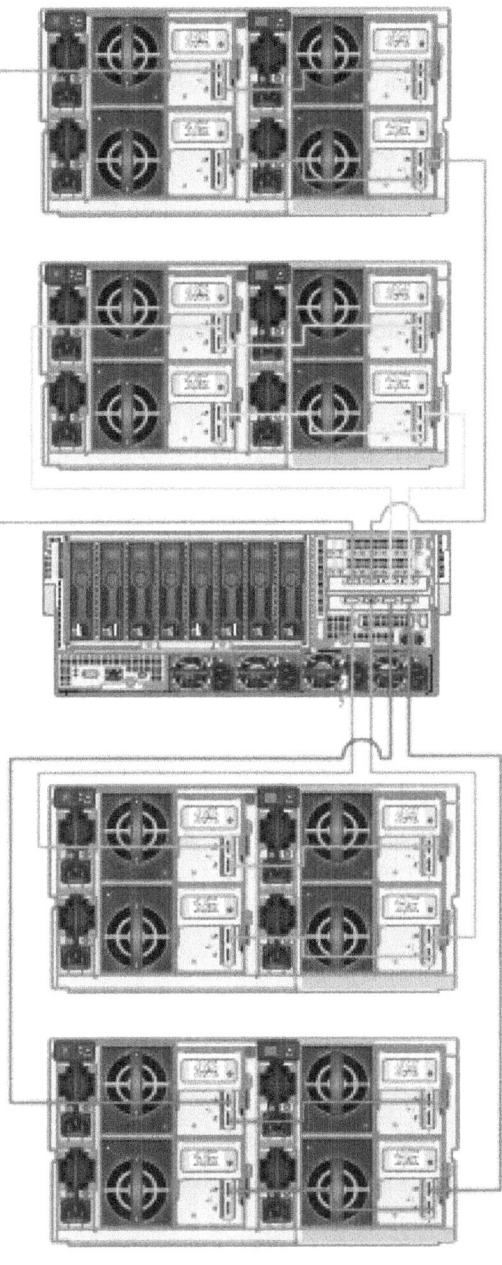

Figure 6-4 Future deployment: Apollo 4510 connected to 4x D6020 enclosures

Figure 6-4 shows the option for future expansion using the D6020's daisy chaining capability. This allows 2x D6020 enclosures to connect to each P841 controller, for a total of 4x D0620 enclosures per Apollo 4510. This diagram shows that drawer-to-drawer connections can be used in order to maintain IO module redundancy. If the first IO module fails in one drawer, the IO traffic will move to the connection through the second drawer. This will accommodate the total future capacity needed for this solution.

In addition to choosing the suitable number of controllers, the following controller settings were selected in order to optimize performance:

1. Align stripe size to the application IO sizes, a good choice for this project since it focuses on storage for a video archive.

2. Because IO performance is not critical in this use case, the Rebuild Priority was set to high. Under this setting, the controller prioritizes and rebuilds over IO requests, with the intention to rebuild failed drives before a second drive can fail. Since our focus is on maintaining the video files, this is a suitable option.

3. We also used Rapid Parity Initialization, which sets the parity bits before data is written to the drives. This delays the original drive initialization, but prevents performance degradation while the data is written.

    There are also options for setting the spare drives in the array:

    - **Dedicated Spare Mode**—In this mode, the spare is assigned based on its slot. If a drive fails, the data will be built on the spare. However, when the original drive is replaced, the new drive will be rebuilt from the data on the spare drive. The spare then returns to acting as spare. Thus, the spare will remain consistently in the same slot. In Dedicated mode, the spare can be shared among multiple RAID groups.
    - **Auto-Replace Mode**—In this mode, the spare moves slots based on rebuilds. If a drive fails, and the data is rebuilt on the spare, the newly replaced drive becomes the new spare. In this mode, the spare cannot be shared among logical drives. In this mode, it is critical to carefully track which slot contains the acting spare.

In this case, we selected Dedicated Spare Mode. This allows you to share spares on each controller, which means more drives can contribute to the total usable pool. This also simplifies the process of tracking the spare, as it is always in the same slot.

Hardware inventory

Table 6-1 shows the Bill of Material (BOM) for a single remote office deployment.

Table 6-1 BOM for remote office solution

Qty	Product #	Description
1	799581-B21	HP Apollo 4510 Gen9 CTO Chassis
1	799581-B21 ABA	U.S. - English localization
4	720479-B21	HP 800W F S Plat Ht Plg Pwr Supply Kit
4	720479-B21 0D1	Factory integrated
1	681254-B21	HP 4U/4.3U Rail Kit
1	681254-B21 0D1	Factory integrated
1	786593-B21	HPE XL450 Gen9 1x Node Svr
1	786593-B21 0D1	Factory integrated
1	847444-L21	HPE XL450 Gen9 E5-2680v3 FIO Kit
1	847444-B21	HPE XL450 Gen9 E5-2680v3 Kit
1	847444-B21 0D1	Factory integrated
16	726719-B21	HP 16GB 2Rx4 PC4-2133P-R Kit
16	726719-B21 0D1	Factory integrated
1	665243-B21	HP Ethernet 10Gb 2P 560FLR-SFP+ Adptr
1	665243-B21 0D1	Factory integrated
2	726821 B21	HP Smart Array P440/4G Controller
2	726821 B21 0D1	Factory integrated
2	726903-B21	HP Smart Array P841/4G controller
2	726903-B21 0D1	Factory integrated
2	652583-B21	HP 600GB 6G SAS 10k 2.5in SC ENT HDD
2	652583-B21 0D1	Factory integrated
68	805337-B21	HPE 8TB 12G SAS 7.2k 3.5in MDL LP HDD
68	805337-B21 0D1	Factory integrated
1	C6N36ABE	HPE Insight Control ML/DL/BL FIO E-LTU
1	761871-B21	HP Smart Array P244br/1G FIO Controller
1	808967-B21	HP Apollo 4510 P440 x2/P840 Cable Kit
1	808967-B21 0D1	Factory integrated
1	727258-B21	HP DL/ML/SL 96W 145mm Smart Stor Battery
1	727258-B21 0D1	Factory integrated
1	799377-B21	HP Apollo 4510 8HDD Rear Cage Kit
1	799377-B21 0D1	Factory integrated
2	P8Y57A	HPE D6020 8TB 12G SAS LFF MDL 560TB Bndl
8	716197-B21	HP Ext 2.0m MiniSAS HD to MiniSAS HD Cbl
4	K2Q23A	HPE D6020 Dual I/O Module Option Kit
2	455883-B21	HP BLc 10G SFP+ SR Transceiver

Implementation project plan

On this project, an HPE Channel Partner was integral to the success of the implementation. The solution components were shipped and assembled at the partner location to ensure that the final solution, when delivered to the client site, was fully operational. Many activities took place at the partner location including installation of the HDD, logical drive setup, installation and setup of management tools, and installation and setup of operating systems. The fully operational solution was then installed at the client location one hundred percent operational.

HPE Insight Control licensing was part of this solution as well, which includes many management components related to server management including HPE Integrated Lights-Out (iLO). This ensures that the IT staff, although not local to the equipment, are able to access the server remotely through iLO's remote console and virtual media capabilities. This is critical because remote IT experts are able to troubleshoot issues from a different location, while generalists perform any required work on-site. Your team can also comfortably manage this remote server deployment using the same management tools as your data center deployments.

Solution validation and success criteria

A variety of tests were performed on this solution to ensure that all of the initial design criteria were met. Much more time-consuming were the HA-related tests. The nature of pulling out components, the rebuilding process, and other activities are, by their nature, more time-consuming.

The following tests were performed to ensure the HA of the solution.

Test 1—Remove one 8 TB drive from a 9+2 RAID 6 set in the D6020 array. The array had one dedicated spare. The drive was approximately 90% full, which is unusually high, considering the enclosure holds 70 drives and data will be striped across the drives in the RAID group. This was done intentionally to provide the highest possible rebuild time and the worst-case scenario. The drive was immediately replaced, and the rebuild was well within the acceptance window.

Test 2—Removed 2x 8 TB drives from a 9+2 RAID 6 set in the D6020 array. In this case, the array had two dedicated spares, and the pulled data drives were not replaced immediately. The data was rebuilt on the two dedicated spares, and after 87.5 hours, the logical drive was recovered. However, since the spares were dedicated, the Smart Storage Administrator did not show the array as recovered until the original two drives are replaced. Once the data was rebuilt on these drives and the original dedicated spares returned to being spares, the logical drive was shown as fully rebuilt. Copying the data from the spares back to the replaced drives took approximately another 28 hours. The data was recovered well within the acceptance window, and the logical array was fully restored after some additional time also within the acceptance window.

Predictive Spare Activation provided by HPE SmartArray controllers begins the rebuild process on a spare drive once a data drive shows signs of failing. This allows for shorter rebuild times after the drive actually fails because the data has begun to copy over already. In these tests, we removed fully functioning drives from the array, meaning that Predictive Spare Activation mode was not used since no signs of failing were noted. This enhanced the "worst-case scenario" testing. Using this tool, you can reduce the rebuild times substantially.

In any rebuild, the hard drives run at highest power/speed during rebuilds to quickly rebuild the new drive from parity, while providing continued IO access. This will cause the drives to wear more quickly. For this reason, there is a higher statistical chance for a second drive to fail once the rebuild of a first failed drive is underway, increasing the chance of data loss. It is important to be aware of this risk when deciding if hardware RAID is an acceptable route for high availability.

Looking ahead

This design meets the current requirements and also has in it future expansion capability. This includes managing the first deployment, expanding at this site, and replicating across different projects and locations.

The following are some of the keys to future growth:

- Daisy chaining creates the ability to add D6020 enclosures for growth. This allows expansion with just a D6020, in just 5U building blocks, when additional storage is needed.

- Management through Insight Control and iLO is key for the Apollo 4510, which is managed in the same manner as other HPE Proliant servers in this environment.

- The design can be used in other locations that have the need for archival of any type of large files. Repetition across locations. This proven design can be quickly replicated to other facilities as required.

Case #2 — Replacing tape with object storage for file archive

This section covers using object storage to eliminate tape backup so that the local files many be accessed more quickly for analysis purposes. Using the same principles as the previous object storage chapter, this solution is used as a target for an existing backup solution. As is the case in many environments, backup software has been implemented in a successful manner and changing the target from tape to object storage greatly enhances the backup environment in a variety of ways. The data now on tape is required for analytics use, and it takes an unacceptable amount of time to get access to the data. Object storage is used in this chapter to solve the problem.

The following list summarizes the problem statement of this project:

- Data on tape needs to be available online for short-term analytics.
- Three PB of data is currently on tape takes a substantial amount of time to read off tape.

- The data archive is growing at 35% annually, which means more data needs to be put on tape every year. The data needs to be retained on tape for only one month.

- The existing tape infrastructure continues to grow and another tape library will need to be added if there is not an alternative way to archive the data and access it for analytics processing.

- Tape access is far too slow and even moderate speed disk access, which can be achieved with object storage, would easily meet the performance requirements.

- The backup solution employs an existing software product for tape backup. This software can be used for the object storage solution eliminating an operational change for backup.

- Private cloud implementation would be ideal to solve the problem and public cloud, as an option, is also desirable.

- Two locations with Active-Passive.

This clear problem statement results in the following success criteria for the project:

- Supports 3 PB storage requirement initially with 35% annual data growth and 35-day retention.

- Supports incremental capacity at 2% daily incremental back-up and a weekly full back-up for 5 weeks.

- Incremental back-up windows is under 6 hours. Full back-up window is under 20 hours.

- The solution has to be sized up for 3 years forward and provide for linear growth with no loss in performance or availability.

- Improved usability and quality of query returns.

- Simplified maintenance.

- The solution must be cost-effective, with a target of below $1 per gigabyte (GB) of storage. This should include all hardware, software, and 3-year support.

- Consider hybrid cloud deployment as a potential future option.

- Successful Active-Passive implementation.

The ideal solution in this case is object storage, which meets all of the requirements including the multisite requirement. Scality RING supports the multisite requirement with this software solution. The following are the key components of the solution:

- Scality Object Storage RING software.

- HPE storage and connector servers.

- Third-party software that supports deduplication and compression software and related hardware infrastructure.

Solution overview

A key component is the two-site mirrored RING (Active-Passive) implementation of the design. The following high-level diagram depicts the solution:

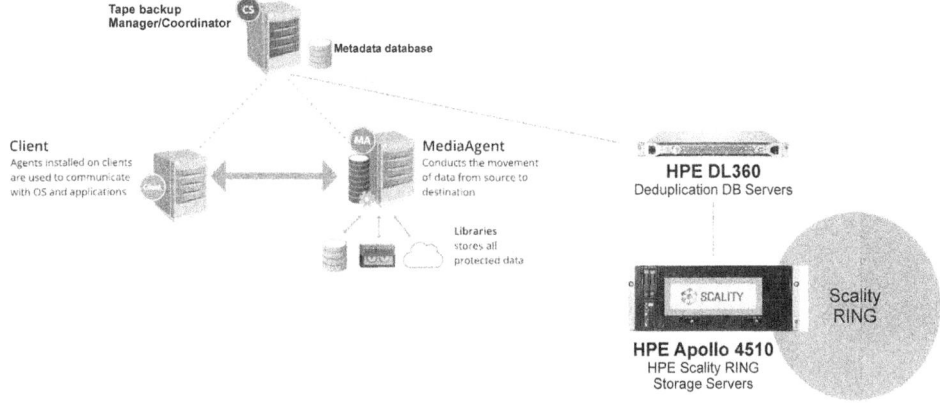

Figure 6-5 High-level solution diagram

The following list describes the aspects of this solution that addresses the requirements:

- **Archive size**—Using current 3 PB storage number and applying growth rates above, in three years, we would be looking at 8 PB usable storage requirement in 3 years. The current backup solution employs deduplication and compression that is included in the capacity calculation. The 8 PB storage calls for 24-Apollo 4510 servers on each site, which are fully populated with the following disks:
    - 5-800GB Serial Attached SCSI (SAS) Solid State Drives (SSD)
    - 63-8TB Serial Advanced Technology Attachment (SATA) Hard Disk Drives (HDD)
- **Data protection**—Data availability is ensured by the Scality RING replication and erasure coding algorithms covered in Chapter 5. In this archive, the average file size is 500 kilobytes (KB), with 90% of all files being larger than 60 KB. Files 60 KB and smaller are protected by triple replication, whereas larger files are protected by erasure coding ARC (9+3). This means that you could lose one data center entirely and up to two nodes in the other data center, and still have access to all the archived data. ARC stands for Advanced Resiliency Configuration, and it is a method of data protection in which data is broken into fragments, expanded, and encoded with redundant data pieces and stored across a set of different locations or storage media.
- **Performance**—Data access is simplified, usability improved, and query performance stabilized by using a standard protocol integrated directly into each storage server. These protocols are supported by the backup software as well.

- **Potential cloud integration**—A standard storage access protocol is used so cloud integration in the future is supported.
- **Simplified maintenance**—Lifecycle management is improved significantly by hardware platform standardization and management tools such as iLO, SPP, HPSUM, and so on. Linux-based Scality software that runs entirely in user space (has no dependency on OS kernel version).
- **Investment protection**—The server environment and storage capacity can be expanded at any time including support of future generations of server hardware. No downtime is required to add servers to this environment.

Solution details

The solution is based on HPE-Scality reference architecture. To support the 3 PB initial storage requirement, the solution was based on the proposed Apollo 4510 servers. There were also DL360 servers used to support the backup software. Below are the details for each component:

- 24 Apollo 4510 servers on each site to support 8 PB of usable object storage with the requisite level of data protection.
- One deduplication database server per 100 TB of data (as recommended by the backup software and used to identify server requirements for future expansion).

Table 6-2 shows the BOM for the Apollo 4510 servers for the project. Note that 24 such servers are needed for each site.

Table 6-2 Apollo 4510 BOM for Tape Elimination Project

Qty	Part number	Description
1	799581-B21	HP Apollo 4510 Gen9 CTO Chassis
4	720479-B21	HP 800W FS Plat Ht Plg Pwr Supply Kit
1	681254-B21	HP 4U/4.3U Rail Kit
1	786593-B21	HPE XL450 Gen9 1x Node Svr
1	783901-L21	HP XL450 Gen9 E5-2630v3 FIO Kit
1	783901-B21	HP XL450 Gen9 E5-2630v3 Kit
12	728629-B21	HP 32GB 2Rx4 PC4-2133P-R Kit
1	665243-B21	HP Ethernet 10Gb 2P 560FLR-SFP+ Adptr
2	726821-B21	HP Smart Array P440/4G Controller
1	758959-B22	HP Legacy FIO Mode Setting
2	655710-B21	HP 1TB 6G SATA 7.2k 2.5in SC MDL HDD
5	797299-B21	HP 800GB 12G SAS VE 3.5in LP EV SSD
63	805334-B21	HP 8TB 6G SATA 7.2k 3.5in MDL LP HDD

Table 6-2 Continued.

Qty	Part number	Description
1	E6U59ABE	HP iLO Adv incl 1yr TS U E-LTU
1	761878-B21	HP H244br FIO Smart HBA
1	808967-B21	HP Apollo 4510 P440 x2/P840 Cable Kit
1	727258-B21	HP DL/ML/SL 96W 145mm Smart Stor Battery
1	799377-B21	HP Apollo 4510 8HDD Rear Cage Kit
2	455883-B21	HP BLc 10G SFP+ SR Transceiver

Table 6-3 shows the BOM for the DB dedup servers for the project. Note that 30 such servers are recommended for 3 PB storage.

Table 6-3 DB dedup server for Tape Elimination project

Qty	Part number	Description
1	755258-B21	HP DL360 Gen9 8SFF CTO Server
1	755386-L21	HP DL360 Gen9 E5-2640v3 FIO Kit
1	755386-B21	HP DL360 Gen9 E5-2640v3 Kit
8	726720-B21	HP 16GB 2Rx4 PC4-2133P-L Kit
1	764630-B21	HP DL360 Gen9 2SFF HDD Kit
10	804625-B21	HP 800GB 6G SATA MU-2 SFF SC SSD
1	764642-B21	HP DL360 Gen9 2P LP PCIe Slot CPU2 Kit
1	665243-B21	HP Ethernet 10Gb 2P 560FLR-SFP+ Adptr
1	QW972A	HP SN1000Q 16Gb 2P FC HBA
1	758959-B22	HP Legacy FIO Mode Setting
1	734807-B21	HP 1U SFF Easy Install Rail Kit
1	843199-B21	HPE Smart Array P840ar/2G Controller
2	455883-B21	HP BLc 10G SFP+ SR Transceiver
2	720478-B21	HP 500W FS Plat Ht Plg Pwr Supply Kit
1	734811-B21	HP 1U CMA for Easy Install Rail Kit
1	E6U59ABE	HP iLO Adv incl 1yr TS U E-LTU
1	843234-B21	HPE DL360 Gen9 P840ar Cable Kit

Summary

The two solutions covered in this chapter offer two different approaches to a similar problem. In the first case, a disk-based backup solution with customer backup software minimized cost and use of floor space while meeting the key requirements of high capacity, availability, and performance. In the second case, the use of object storage was a key design factor, with the Scality RING architecture supporting two sites that could also be expanded to include additional sites and support for cloud integration. From a workload perspective, this solution provided a great improvement over the performance of the tape drives while meeting other success criteria for streamlined IT operations such as simplified maintenance and a flexible environment for future deployments.

7 Hosted Desktop Infrastructure (HDI)

INTRODUCTION

In the Financial Services Industry (FSI), use of trader workstations is at the heart of daily trading and requires dedicated hardware that supports high speed access to applications. The trader workstation itself typically consists of multiple displays showing various applications such as data, spreadsheets, analytical applications, and other FSI business applications. Access to applications must have high and predictable performance and this must be achieved at an acceptable cost.

The workload design goals included the following customer requirements:

- **Performance**. High and predictable performance is essential. As close to real-time access to information as possible is required in order for the trader workstation to be an effective tool.
- **Physical security**. The hardware supporting the trader workstation must be located in the data center, in an environment that is protected from unauthorized access. In addition, the use of dedicated hardware minimizes the risk of drift (unauthorized configuration and OS changes.), something not always feasible in virtualized environments.
- **Management**. The environment must be easy to set up, manage and maintain and be able initially to support a total of forty-five trader workstations.
- **Cost**. Total cost of ownership (TCO) considerations range from the price of hardware and speed of deployment to the operational costs over the solution's life cycle, such as the use of power and space.

The HPE Moonshot 1500 chassis provided the ideal platform to meet these requirements, providing a Hosted Desktop Infrastructure (HDI) solution that would assure predictable performance and provide a non-shared environment dedicated to the desktop.

Solution overview

At the center of the solution is the Moonshot 1500 chassis with 45 m710p cartridges. This is an HDI solution with high-performance capacities. The hardware standardizes the use of low-power processors in cartridges that plug into the chassis. The cartridges are workload-optimized compute engines

CHAPTER 7
Hosted Desktop Infrastructure

with common management and networking services. Each remote user is assigned one cartridge. The chassis itself is 4.3U and multiple chassis can be deployed per rack.

 **Note**

There are multiple, workload-optimized cartridges, based on different processors. This design uses the m710p cartridge that has one processor per cartridge with four cores. Table 7-1, showing the full m710 family, appears later in the chapter.

Integrated Lights-Out (iLO) is used for controlling and monitoring Moonshot. Each chassis requires only a single iLO IP address for all the cartridges. Thus, deploying the entire chassis requires making an iLO connection, the switch uplink modules network connections, and power connections. An entire chassis can be brought online in a matter of minutes. Cartridges are connected to the internal switches through connections inside of the Moonshot 1500 chassis.

In this solution, thick clients are used, running Citrix Receiver. The Moonshot cartridges run Windows 7 with Citrix Xen Desktop Virtual Delivery Agent (VDA) 7. The remote user has 3–4 active screens. One m710p cartridge is deployed per remote user.

With the Moonshot design each cartridge is connected to the two internal 45XGC switches. Each cartridge has a dual ported 10GbE NIC. Each switch has an Uplink Module for connecting the switch to the network, saving customer network ports. The Uplink Modules can be 4x 40GbE or 16x 10GbE. Figure 7-1 depicts the networking scheme:

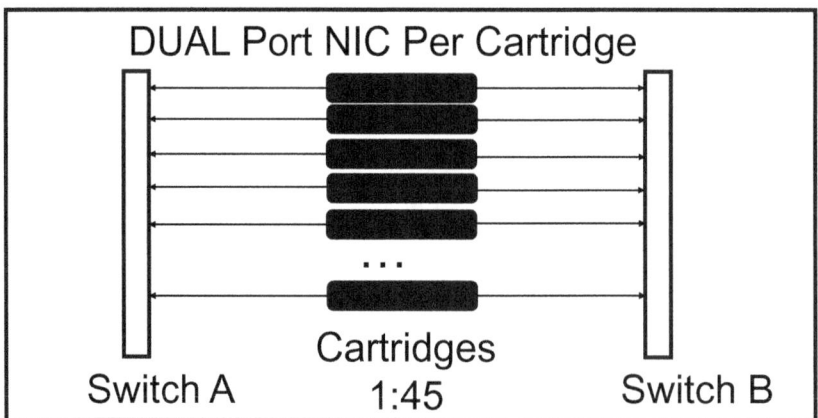

Figure 7-1 Cartridge-to-switch connections

Figure 7-2 depicts the overall solution.

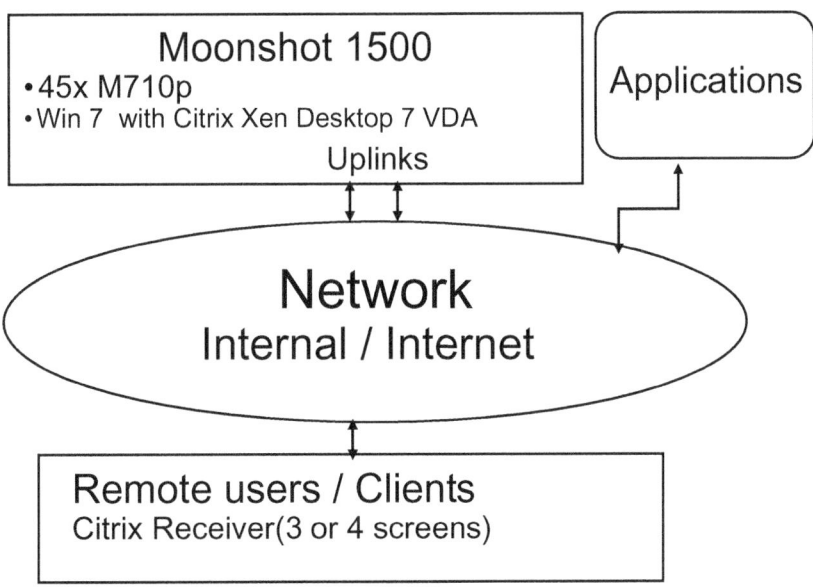

Figure 7-2 HDI solution overview

Figure 7-2 shows the Moonshot solution and applications at the top, the internal network for connectivity, and then the remote users and clients at the bottom.

Moonshot is a different form factor from other enclosures. Figure 7-3 shows the Moonshot chassis, cartridges on the left, and a diagram of the back of the chassis on the right.

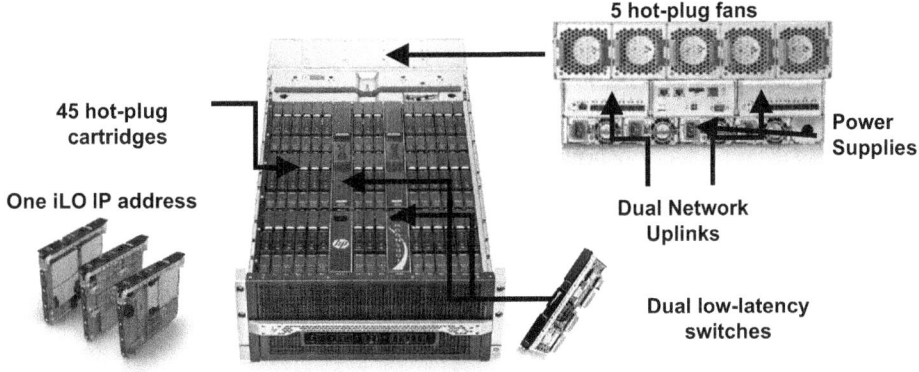

Figure 7-3 Moonshot components

As noted earlier, the solution uses the m710p and Table 7-1 shows the full M710 family of cartridges at the time of this writing:

Table 7-1 m710 Cartridge options

Cartridge	Processor	GPU	NIC	Memory	Storage	Density
m710	Xeon E3-1284Lv3, 4 core	Iris PRO 5100	Dual Port 10GbE/ CPU	32GB	Local SSD (NGFF/ m.2)	45/4.3U
m710p	Xeon E3-1284Lv4, 4 core	Iris PRO P6300	Dual Port 10GbE/ CPU	32GB	Local SSD (NGFF/ m.2)	45/4.3U
m710x	Xeon E3-1585Lv5, 4 core	Iris PRO P580	Dual Port 10GbE/ CPU	64GB	Local SSD (NGFF/ m.2) 4TB	45/4.3U

How did we arrive at this solution?

The Moonshot solution was designed with the following key considerations in mind:

- Predictable performance—Solutions based on virtualization of physical servers or blades are inherently more complex to size, tune, and manage. Conversely, using physical infrastructure can be prohibitively expensive to purchase and operate. The Moonshot cartridge solution is low cost, power space efficient, and easy to manage. There are many other Moonshot cartridges and a variety of processors that are optimized for different workloads.

- Data center space and power optimization—The Moonshot solution is space efficient in that the forty-five cartridges in this design can be accommodated in 4.3U. Each m710p cartridge is devoted to one user (serving 45 users in total).

- Management—Moonshot is managed on a chassis-basis and is very quickly deployed. The chassis has a Chassis Management Module that is used to configure and control the chassis. A single IP address is needed to manage iLO for all cartridges. The m710p provides Command Line Interface (CLI) access, but its successor, the m710x and other new cartridges, allows Graphical User Interface (GUI) iLO access via the Chassis Management module. Moonshot can be administered on a chassis basis, resulting in operational efficiencies.

- High availability—If a cartridge fails, the user can be moved onto another cartridge. This is a manual process that was implemented with a script in this installation.

- Security—The Moonshot solution is physically secure because it can be housed in the data center. Since this particular solution also allows user access across the Internet, it relies on the customer's DMZ security. Security is further enhanced if running Windows 10, with its added security features, and Universal Extensible Firmware Interface (UEFI). Although this solution is based on the m710p, the successor m710x provides UEFI features such as Secure Boot, which is designed to stop malicious code from corrupting the OS boot Loader and causing havoc. The m710x also supports Trusted Platform Module (TPM) 2.0 consistent with ProLiant servers, and the iLO4 security features such as LDAP and SSL encryption. Similarly, using Windows 10, instead of Windows 7, also enhances security. Windows 10 security feature enhancements are described in the following Microsoft article: https://technet.microsoft.com/en-us/itpro/windows/keep-secure/windows-10-security-guide

- Cost—TCO is enhanced by the simplicity of installation, power and space savings, simplified operations, and efficient management over the solution's life cycle.

Solution inventory

Digging into the details of this solution, you will see that the configuration consists of the key components of a Moonshot 1500 chassis with 2x 45XGC (45x 10GbE ports) and 45x m710p cartridges. Dual switches are used in this solution to enhance network bandwidth and availability. Each cartridge is prewired within the chassis to access both switches. Each cartridge has 2x 10GbE ports wired to the two switches, and these network ports can be combined or "teamed."

The 45XGC switch has the following characteristics for each switch:

- 45x 10GBE ports downlinks to the cartridges.
- An Uplink Module, protruding through the rear and connects the switch to the network.
- There are two options for Uplink Modules: either 4x 40GbE ports (4 QSFP+) or 16x 10GbE (16xSFP+)
- The choice of Uplink Modules depends on your existing networking infrastructure. There are also multiple cabling options

Table 7-2 lists the hardware components for this solution.

 **Note**
Cables may vary based on the installation and are not listed.

Table 7-2 Hardware Components for HDI solution

Qty	Part#	Description
1	755372-B21	HP Moonshot 1500 Chassis Opt OS
1	ZU707A	HPE Enclosure Customization Package
1	HA841A1	HPE Customer Supplied Asset Tag SVC
4	684532-B21	HP 1500W Ht Plg Pwr Supply Kit
4	684532-B21 0D1	Factory integrated
2	704652-B21	HP Moonshot-4QSFP+ Uplink Module Kit
2	704652-B21 0D1	Factory integrated
2	704654-B21	HP Moonshot-45XGc Switch Kit
2	704654-B21 0D1	Factory integrated
1	681254-B21	HP 4U/4.3U Rail Kit
1	681254-B21 0D1	Factory integrated
1	681260-B21	HP 0.66U Spacer Blank Kit
1	681260-B21 0D1	Factory integrated
45	808915-B21	HP ProLiant m710p server cartridge
45	808915-B21 0D1	Factory integrated
45	765483-B21	HP Moonshot 480G SATA VE M.2 2280FIO Kit

Client configuration

A variety of different options can be deployed including thin client or a full-function desktop. In this case, we deployed the following personal computer:

- A PC with i7 processor
- 16GB RAM
- SW: Windows 7 and Citrix Receiver (4.5 for Windows)
- Three or four screens

Cartridge configuration

As mentioned earlier, the m710p cartridge is used in this solution consisting of the following:

- m710p with 1x processor, E3-1284L v4 4 core, 32GB RAM
- Software is Windows 7/Citrix Xen Desktop Virtual Delivery Agent (VDA) 7.6.300
- 480GB SSD

> **BIOS Note**
> ProLiant Gen9 servers support Unified Extensible Firmware Interface (UEFI) class 2 (both UEFI and Legacy BIOS boot modes.) Moonshot cartridges support either UEFI or Legacy BIOS mode. The m710p supports legacy BIOS, while the m710x is UEFI-based. Here are some considerations related to BIOS and operating systems:

- Windows 10 and other operating systems offer secure boot, which is a desirable feature that requires UEFI. Windows 10 can run on legacy BIOS, but the UEFI dependent security features are not available.
- Preboot Execution Environment (PXE) is used to boot servers in many installations. UEFI environments may require networking configuration changes before PXE can be used. This is part of the planning exercise.

Citrix software

There are two Citrix components used with this solution.

- The Citrix Receiver, which runs on the client system.
- The Citrix Xen Desktop Virtual Delivery Agent (VDA) runs on Microsoft Windows in the Moonshot cartridge.

Many organizations use and have access to Citrix software, including the two listed above, so this software was not part of the HPE proposal in this solution.

Moonshot software

The solution relies on installing the most current Moonshot software and firmware that includes a bundle called the Moonshot Component Pack (MCP.) The latest MCP 2016.07.0 can be downloaded from the HPE web http://h17007.www1.hpe.com/us/en/enterprise/servers/products/moonshot/component-pack/index.aspx.

MCP is a comprehensive firmware solution tested on Moonshot and delivered as a compressed file. The compressed file includes all the component files needed to update a Moonshot system. You can deploy the firmware updates contained in the MCP through the iLO Chassis Manager Command Line Interface (CLI) or HPE Moonshot Switch Module CLI. This can be accomplished using HP Smart Update Manager (SUM), which is included with the files, or manually installed. The latest pack can be downloaded at http://www.hpe.com/info/moonshot/download. The MCP is the Moonshot equivalent of the Support Pack for ProLiant (SPP) for servers.

Moonshot power supplies

The Moonshot chassis are installed in racks in the data center. Each contains four HPE FlexPower, power supplies. In a typical configuration, N+N power redundancy is desired. A fully loaded chassis will typically consume 2.5 KW–3 KW. The number of chassis that can be supported per rack will depend on the data center power and cooling capabilities and on the maximum wattage allowable per rack at a given data center.

Note
Power consumption will vary depending on the specific components and can be modeled using the HPE Power Advisor tool.

System management

The solution is managed via the Moonshot Chassis Management module using the Moonshot iLO interface that is different from the iLO on ProLiant servers. As mentioned earlier, the Chassis Management Module CLI is used to manage the chassis. By design, the chassis is designed to be managed as a single unit.

Note
HPE also provides the Cluster Management Utility (CMU), https://www.hpe.com/us/en/product-catalog/detail/pip.hpe-insight-cluster-management-utility.3296361.html, which is another consideration if your environment has a lot of components that should be managed as a cluster.

Summary

This solution is working in production with traders and has resulted in the following benefits:

- High and predictable performance
- Rapid installation and onboarding of new users
- Efficient management
- Reduced power and cooling costs
- Greater density that reduces data center space
- Management simplicity through virtualization avoidance

In addition to this successful trader workstation HDI-based solution, there are many other HDI and VDI solutions that are described under HPE Moonshot for Mobile Workspace solutions at the following link: http://www8.hp.com/us/en/products/servers/moonshot/mobile.html

8 Virtual Desktop Infrastructure (VDI)

INTRODUCTION

Client Virtualization in the form of Virtual Desktop Infrastructure (VDI) has become a mainstream solution for delivering a physical desktop experience for large and diverse populations of users. VDI provides a datacenter-based ecosystem that is inherently more secure and resilient than the physical desktop environment. In addition, user productivity increases due to faster system startup and minimized time to recover desktop access when issues arise contributes to overall organizational efficiency.

Our VDI workload design goals meet the following generic customer requirements:

- A broad range of users
- Host hundreds to thousands of total users with a "building block" approach
- High-availability as part of the design
- Ease transition from Windows 7 to Windows 10
- Easy to implement and expand
- Simple solution-level management
- Minimal capital expenditure
- Low Total Cost of Ownership (TCO)

Solution overview

This solution, based on the HPE HyperConverged 380 (HC380 in this chapter) and HPE StoreVirtual VSA, is ideal to meet our stated design goals. As with all such designs, we performed an assessment, gathering empirical data of the existing environment, to craft this design.

The HC380 supports a variety of processors and RAM capacity that will vary based on specific design needs for a solution.

Depicted in Figure 8-1 is the high-level functional architecture for this HC380 VDI solution.

CHAPTER 8
Virtual Desktop Infrastructure

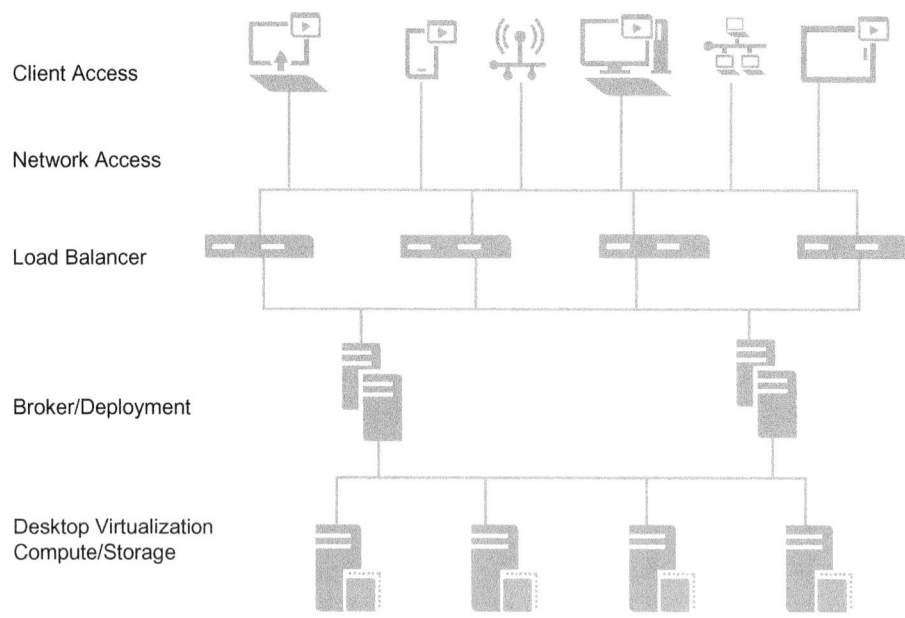

Figure 8-1 VDI diagram

This VDI workload solution consists of the following hardware components:

- 16-node HC380 cluster:
    - 16 DL380 servers
        - Dual Intel E5-2697v4 processors
        - 1024 GB RAM
        - 2 X 13.4TB Storage Blocks
        - Dual 10GbE
- Top of Rack Switches
    - 2 X 5940 48XGT—Storage and general access
    - 5130 24G—ILO and Management
- Software stack covered later
- N + 1 Server High Availability (HA) which is full capacity and performance of the cluster is maintained when a single server node is removed for service

These components are shown in Figure 8-2:

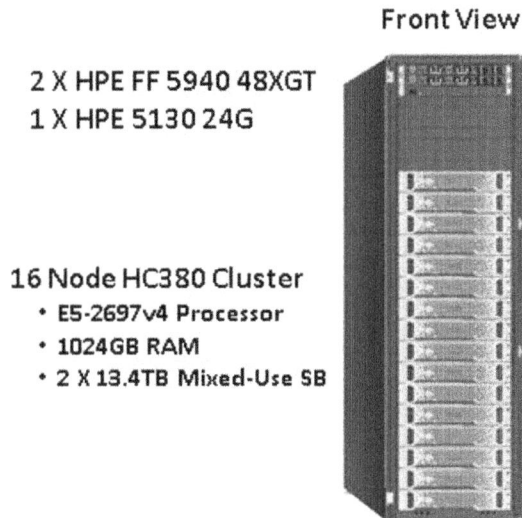

Figure 8-2 HC380 VDI rack diagram

How did we arrive at this solution?

This installation had been an early adopter of VDI technologies and has an extensive user population already deployed. The main drawback to the existing VDI environment was a restrictive design that created some major issues in providing HA and disaster recovery on an enterprise scale.

Another challenge was the increasing cost to IT of managing their growing need for more VDI desktops, as well as the impact of new user functionality including collaboration and IP phone. Finally, a migration from Windows 7 to Windows 10 had to be factored into sizing the solution.

Sizing a VDI solution is a matter of optimizing the processor speed and core count, physical RAM capacity, storage capacity, and storage performance to provide the optimal user density per server that provides the best possible user experience.

First, the user experience expectations must be thoroughly understood in order to model an appropriate virtual desktop configuration. One best practice is completing a user activity assessment to capture utilization data over a period of weeks for a representative subset of the total user population. With this data, we develop a model virtual desktop configuration that will aid in identifying the underlying server and storage needs.

In this case, there was an active VDI environment that provided a desktop configuration target for sizing our solution. This configuration consisted of two virtual CPUs (vCPU), 8 GB RAM, and 120 GB persistent desktop for 90%–95% of their user population. The remaining 5%-10% required much more vCPU, RAM, and storage. We accommodated these users by increasing server count appropriately to meet the expanded resource requirements within an individual cluster.

CHAPTER 8
Virtual Desktop Infrastructure

Table 8-1 Customer requirements

Customer requirement	Details
Users	1200–1500
Operating system	WIN7 today. WIN10 in near future.
RAM per desktop	8 GB (90%); 16–32 GB (10%)
vCPU per desktop	2 (90%); 4–8 (10%)
Desktop image size	> = 120 GB
Persistent/non-persistent	100% Persistent today: move to non-persistent in future
Graphics (GPU acceleration)	None — Today
Storage type	All flash

Processor selection

Selecting the correct processor for VDI involves matching the processor speed with number of cores available for an optimal thread-per-core oversubscription ratio. With VDI, we do not consider hyper-thread cores when calculating this ratio and instead use physical cores.

An 8:1 ratio is a good starting point for a generic workload sizing exercise. This "generic" desktop is considered to be 2 vCPU, 4 GB RAM, 60 GB desktop image, 20 IOPS.

However, as processor speed and core counts increase, a more aggressive 10:1 ratio may work for many VDI workloads. In some cases, VDI workloads at 11:1 on the latest Xeon v4 processors have been used, but this is highly user-dependent. Table 8-2 shows the processor options considered at the time of this writing.

Table 8-2 VDI processor options

Standard VDI processor options			
Processor		Clock	Cores
Xeon E5-2680 v4	Broadwell	2.40 GHz	14
Xeon E5-2690 v4	Broadwell	2.60 GHz	14
Xeon E5-2697A v4	Broadwell	2.60 GHz	16

To obtain the greatest number of users-per-server, we need to consider core count balanced with processor speed. As seen in Table 8-2, we narrowed our processor selection pool to processors with highest core count at 2.40 Ghz–2.60 Ghz, given that the majority of users would not need a higher speed processor.

In some instances, such as 3D CAD engineering desktops or other applications with a high performance requirements, a 3.0 GHz or higher processor is needed due to the compute-intense nature of the user experience. In these instances, the E5-2687W v4, 3.0 GHz, 12-core processor could be necessary; however, there would be an associated reduction in the number of users-per-server due to a reduction in the core count.

StoreVirtual VSA and Hypervisor Compute Considerations

The thread-per-core ratio also has the compute needs of VSA and the hypervisor requirements. Roughly, five cores-per-server are estimated to handle VSA and the hypervisor, primarily to handle the increased load associated with SSD I/O in the server and storage communication with the other nodes in the cluster.

What this means for users-per-server calculation is a reduction in the overall potential vCPU total on the server. For this particular workload, we selected the E5-2697Av4 processor to give us a total of 32 physical cores. From this 32 we reserved 4, leaving us with 28 cores to host VDI vCPUs.

Table 8-3 Available processor cores

Core counts after VSA + Hypervisor reservations			
2 Physical processors		Total cores	Available cores
Xeon E5-2680 v4	Broadwell	28	24
Xeon E5-2690 v4	Broadwell	28	24
Xeon E5-2697A v4	Broadwell	32	28

Once we have identified the number of physical cores, we can apply the thread-per-core ratio to determine how many vCPUs can be hosted. In the case of the E5-2697v4 with an 8:1 ratio, we can host approximately 100 desktops depending on the desktop mix and hypervisor overhead.

We typically like to model the solution based on the assumption of a single, homogenous workload throughout the cluster. In this particular case, we needed to host a collection of different VDI desktop configurations that could impact the user-per-server capacity.

StoreVirtual VSA performance and capacity

In addition to delivering the compute and RAM capacity for the desktops, the HC380 StoreVirtual VSA storage capacity and performance in the form of aggregate Input/Output Operations (IOPS) for the HC380 appliance must be determined.

Note that the since we have designed in a single server node for HA capacity, we must ensure that the 15 node cluster provides 196 TB storage, which meets the minimum 192 TB capacity required for this solution.

Table 8-4 identifies the 15 node storage characteristics based on using 2 all-flash storage blocks (SB) in each server node.

CHAPTER 8
Virtual Desktop Infrastructure

Table 8-4 HC380 storage capacity and performance

Total nodes	Storage block	Blocks per node	Total usable storage	Total estimated IOPS
15	13.4 TB SSD	2	196 TB	244600

We now have enough information to generate an appropriate Bill of Materials (covered in the next section).

Solution inventory

Table 8-5 shows the Bill of Materials (BOM) for this VDI workload for the 16-node HC380 cluster that provides for a single server to be removed from the cluster without impacting user capacity or user experience performance (N + 1 Server HA).

Table 8-5 BOM for HC380 VDI solution

Item	Product Number	Option	Description	Quantity
0100	H6J68A		HPE 42U 600x1200mm Advanced Shock Rack	1
	H6J68A	001	HP Factory Express Base Racking Service	1
			HC 380 Appliance for VDI [#1] 16 nodes	
0200	P9D74A		HPE HC380 Cluster Node	16
	P9D74A	0D1	Factory integrated	16
	P9D74A	002	HPE HC380 VDI SW	16
0201	817955-L21		HPE DL380 Gen9 E5-2697Av4 FIO Kit	16
0202	817955-B21		HPE DL380 Gen9 ES-2697Av4 Kit	16
	817955-B21	0D1	Factory integrated	16
0203	805358-B21		HPE 64GB 4Rx4 PC4-2400T-L Kit	256
	805358-B21	0D1	Factory integrated	256
0204	817011-B21		HP 1.92TB 6G SATA MU-3 SFF SC SSD	256
	817011-B21	0D1	Factory integrated	256
0205	749974-B21		HP Smart Array P440ar/2G FIO Controller	16
0206	726897-B21		HP Smart Array P840/4G Controller	16
	726897-B21	0D1	Factory integrated	16
0207	783009-B21		HP DL380 Gen9 8SFF SAS Cable Kit	32
	783009-B21	0D1	Factory integrated	32
0208	786092-B21		HP DL380 Gen9 8SFF H240 Cable Kit	16
	786092-B21	0D1	Factory integrated	16

Table 8-5 Continued.

Item	Product Number	Option	Description	Quantity
0209	700699-B21		HP Ethernet 10Gb 2P 561FLR-T Adptr	16
	700699-B21	0D1	Factory integrated	16
0210	720479-B21		HPE 800W FS Plat Ht Plg Pwr Supply Kit	32
	720479-B21	0D1	Factory integrated	32
0211	768900-B21		HP DL380 Gen9 Sys Insght Dsply Kit	16
	768900-B21	0D1	Factory integrated	16
0212	719073-B21		HP DL380 Gen9 Secondary Riser	16
	719073-B21	0D1	Factory integrated	16
0213	733660-B21		HP 2U SFF Easy Install Rail Kit	16
	733660-B21	0D1	Factory integrated	16
0214	666988-B21		HP 2U Security Bezel Kit	16
	666988-B21	0D1	Factory integrated	16
0215	733664-B21		HP 2U CMA for Easy Install Rail Kit	16
	733664-B21	0D1	Factory integrated	16
0216	758959-B22		HP Legacy FIO Mode Setting	16
0217	P9D85A		HPE HC380 Base SW Image 6.0 FIO Kit	16
0300	H1K92A5		HPE 5Y Proactive Care 24x7 Service	1
	H1K92A5	XW4	HPE HC380 Cluster Node Support	16
0400	BW932A		HPE 600mm Rack Stabilizer Kit	1
	BW932A	B01	Include with complete system	1
0500	AF528A		HP 5xc13 PDU Extension Bars Kit	2
	AF528A	0D1	Factory integrated	2
0600	AF525A		HPE Intelligent 7.3kVA/60309/INTL PDU	2
	AF525A	0D1	Factory integrated	2
0700	HA113A1		HPE Installation Service	1
	HA113A1	5BY	HPE Rack and Rack Options Install SVC	1
0800	P1F60A		HPE Inside Unpack/Clean Up XL SVC	1
0900	HA124A1		HP Technical Installation Startup SVC	1
	HA124A1	5Z0	HPE HyperConverged 380 Startup SVC	16
1000	JH394A		HPE FF 5940 48XGT 6QSFP+ Switch	2
1100	H1K92A5		HPE 5Y Proactive Care 24x7 Service	1
	H1K92A5	ZXP	HPE 5940 Fixed 48G Support	2
1200	HA114A1		HP Installation and Startup Service	1

Table 8-5 Continued.

Item	Product Number	Option	Description	Quantity
	HA114A1	5RN	HPE Top of Rack Startup SVC	2
1300	JC680A		HPE 58x0AF 650W AC Power Supply	4
	JC680A	B2C	JmpCbl-ROW	4
1400	JG552A		HPE X711 Frt(prt) Bck(pwr) HV Fan Tray	4
1500	JG932A		HPE 5130 24G 4SFP+ EI Switch	1
	JG932A	ACD	Switzerland - English localization	1
1600	H1K92A5		HPE 5Y Proactive Care 24x7 Service	1
	H1K92A5	TER	HPE 5130 24G EI Switch Support	1
1700	HA114A1		HP Installation and Startup Service	1
	HA114A1	5Q6	HPE Networks 36/51 Install Startup SVC	1
1800	AF595A		HP 3.0M,Blue,CAT6 STP,Cable Data	48

Software for VDI environment

The HC380 solution provides the basic hypervisor (VMware ESXi) infrastructure because there are several modules of software that are essential to the operation of this VDI solution shown in Table 8-6.

Table 8-6 Software used to implement this VDI environment

Item #	Product description	Additional details
1	VMware vSphere 6.0	ESXi virtualization OS
2	Citrix XenDesktop	Citrix VDI ecosystem
3	User Productivity Apps	Identified by Use Case

The software listed is commonly used in VDI implementations. The specific user applications vary based on the needs of users in the specific implementation.

Summary

As you can see from the preceding explanation, there were several steps and tools involved in crafting this design. The assessment is a key component in gathering important data about the existing environment and user experience. The tools help ensure that the sizing is calculated based on many previous designs and best practices. There were also many experts involved who know not only VDI but also the compute, storage, software, networking, and other components.

9 engineering Virtual Desktop Infrastructure (eVDI)

INTRODUCTION TO eVDI WORKLOAD

Supporting an engineering environment with traditional workstations can be a challenge. Unlike a traditional VDI, client engineers have significantly more resource demands on their workstation. This includes compute, memory, and graphics processing for graphic-intensive workloads. The solution covered in this chapter uses Graphics Processing Unit (GPU) as a key part of the solution.

The engineering Virtual Desktop Infrastructure (eVDI) addresses such challenges by centralizing the computer resources in a data center, located in close proximity to high-speed networks, storage, backup, and other services they require. An additional benefit is providing a controlled and monitored environment for Heating, Ventilation, and Air Conditioning (HVAC,) redundant power, and higher levels of physical security that is often required.

Like many eVDI solutions, the existing environment was outdated, and new technology was required. This created the desire for a new solution that is more agile, scalable, and secure.

Why eVDI

The eVDI workstation solution reduces hardware acquisition costs and improves the efficiency and effectiveness of the engineers. eVDI reduces the need to purchase multiple workstations for engineers who are working on multiple programs.

This solution is based on the HPE BLc7000 Blade Infrastructure, the WS460c-Gen9 PCI Expansion blade providing support for multiple GPUs and a 3Par all 8450 array.

Business benefits of eVDI

There are a variety of benefits to eVDI which are listed below:

1. **Collaboration**
   a. Eliminate data replication and islands of data
   b. Enable Subject Matter Expert (SME) access to critical work as needed

2. **Business process**
   a. Enable access to multiple environments from single productivity unit
   b. Support any tool without multiple physical computers
   c. Enable critical SME access to support any client or issue
   d. Meet business needs with flexible work environment
3. **Security**
   a. Data resides on SAN and is protected in data center and not on local devices
   b. Data is not at risk in transit or on local devices
   c. Virtual machines can be further protected with Internet access and e-mail access removed
   d. Nonemployee (contractors, joint ventures, consultants, and so on) do not "take" any data during their engagement
4. **Mobility**
   a. Provides the option for Bring Your Own Device (BYOD)
   b. Work anywhere, anytime
   c. Any device can have workstation performance
5. **Asset utilization**
   a. Higher utilization of engineering class equipment
   b. Enables sharing of more assets like licenses
   c. Rapid deployment and access to modern tools

These benefits are part of the solution overview in the next section.

Solution overview

This eVDI workload was specifically designed to meet the demands of an aerospace engineering organization with the following requirements:

- Provide a scale-out architecture to support >100+ eVDI users
- High-availability as part of the design
- Support for multiple NVIDIA® Tesla® M6 GPUs options
- 100 GB of storage to be provided per desktop
- Gold/Master desktop images to be provided and updated semiannually

- Installed operating system will be Windows 7 x64
- Easy to implement and expand (Building Block design)
- Simple solution-level management

Figure 9-1 depicts this solution at a high-level.

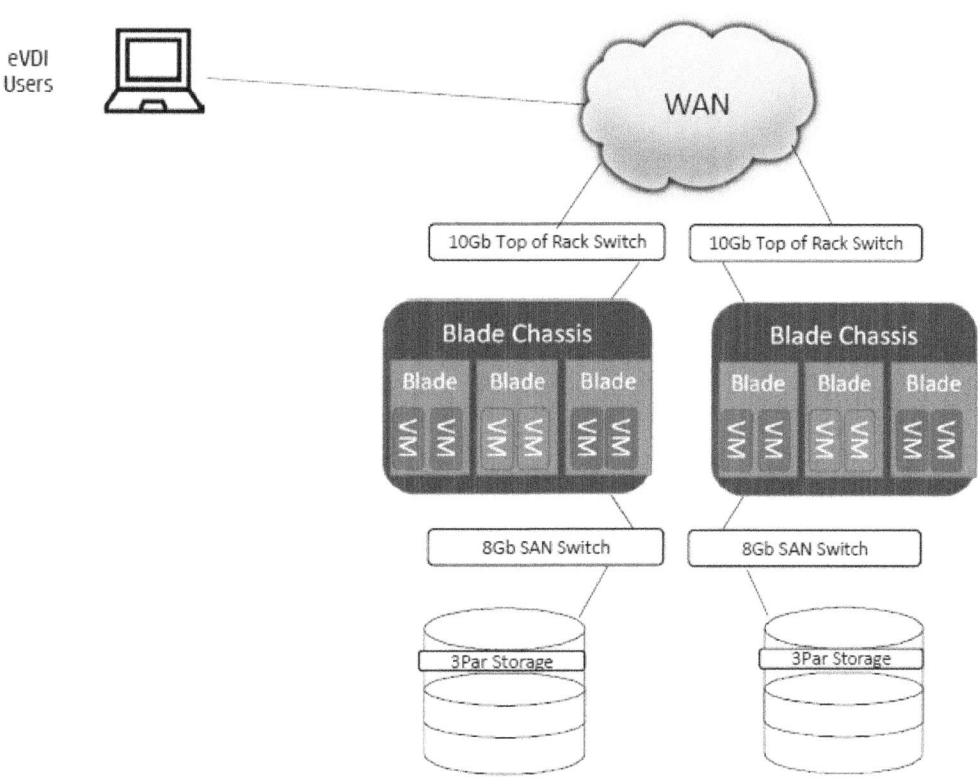

Figure 9-1 High-level overview

CHAPTER 9
engineering Virtual Desktop Infrastructure

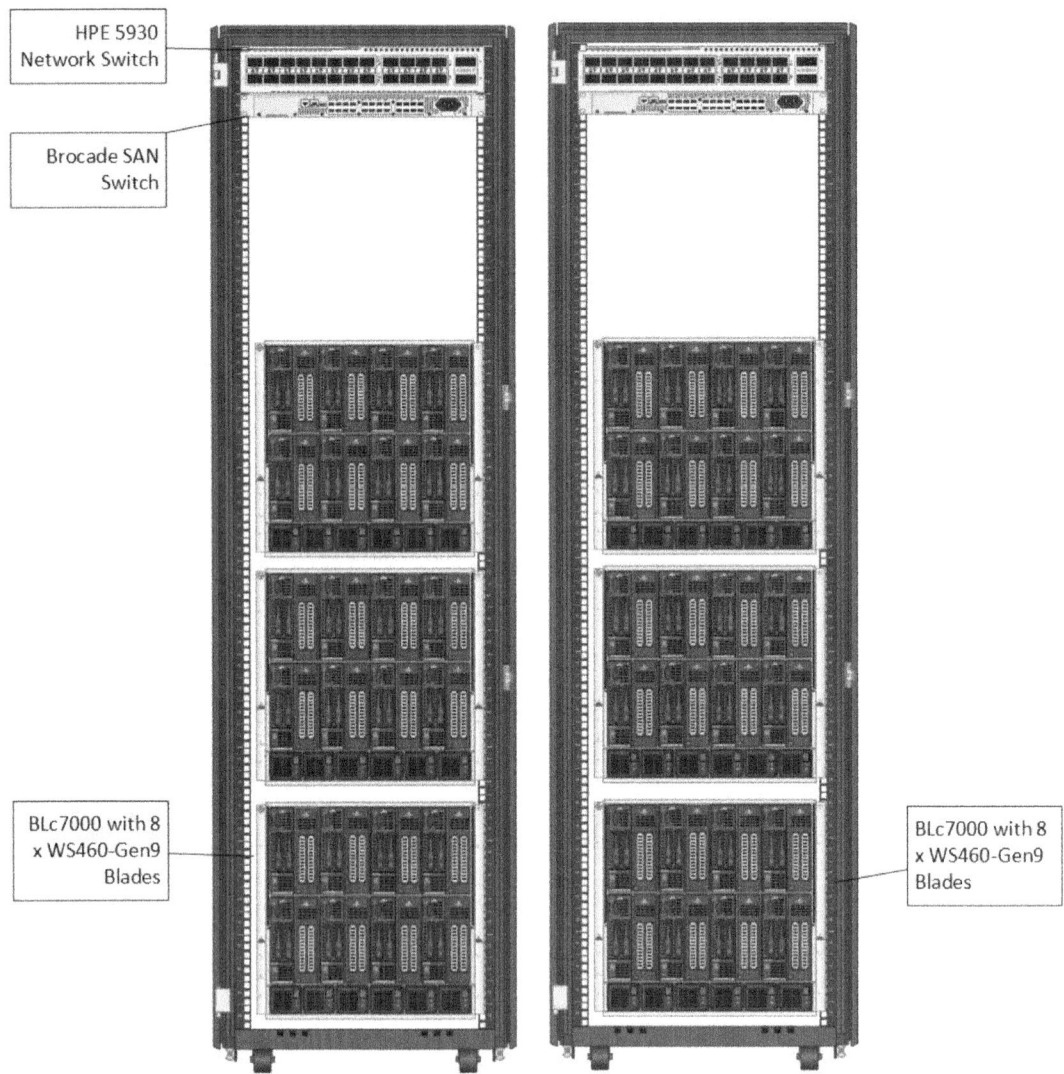

Figure 9-2 Scale-out rack view of the compute building block

Figures 9-2 and 9-3 depict the rack diagrams of the eVDI solution. Figure 9-2 shows support for 96 eVDI users per Rack.

TRIED AND TESTED SOLUTIONS TO ACCELERATE IT AND BUSINESS TRANSFORMATION
Ideal Platforms for Optimizing IT Workloads

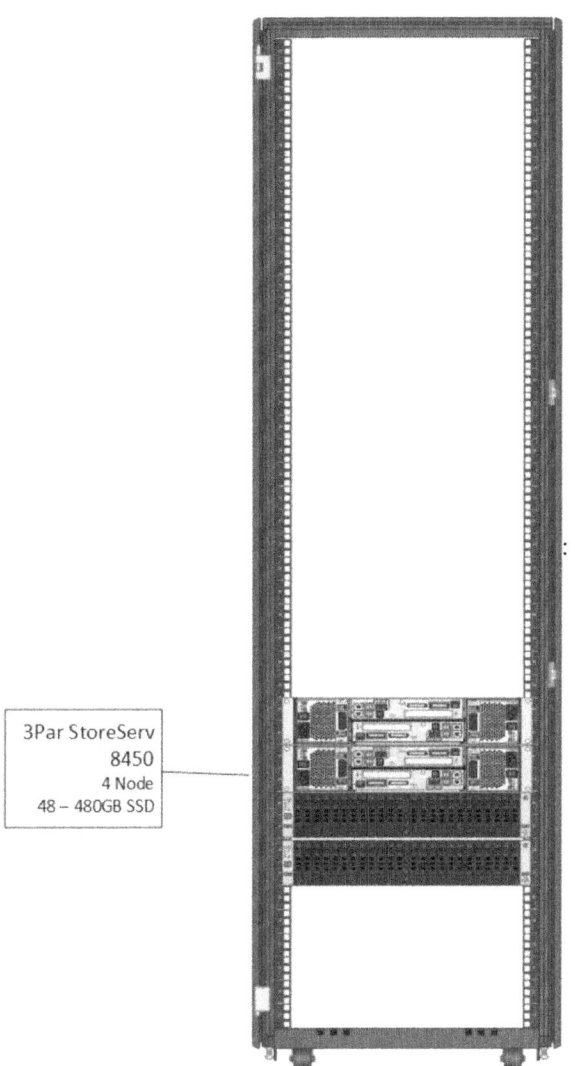

Figure 9-3 Rack view of the 3par Storage

This eVDI solution consists of the following hardware components:

- 1 BLc7000 Blade Enclosures:

    WS460-Gen9 Expansion Blade

    Dual Intel E5-2667v4 8C 3.2GHz processors

    384 GB RAM (12 x 32GB)

    2 X 120GB SSD Drives

> Dual 20Gb FlexFabric adapter
>
> 2 GPUs provided by included NVIDIA® Tesla® M6 cards
>
> NVIDA Tesla perpetual Workstation licenses

Redundant power

Redundant fans

Redundant Onboard Administrators

Redundant Virtual Connect Interconnect Modules

 Note

Only six DIMMs are supported per socket with processors exceeding 135 Watts, so careful consideration is needed with CPU and memory selection to arrive at the target requirement. As the next generation of HPE servers become available, additional combinations of CPU and memory should be considered.

The NVIDIA® Tesla® GPU accelerator was selected as it was optimized for the HPE blade servers and delivers high-performance graphics to virtual workstations allowing up to four GPUs to be installed in an HPE WS460-Gen9 blade. With the HPE WS460-Gen9 expansion blade, the GPUs are added as an option. Coming in a standard PCIe form factor and installed in the 2 PCI expansion slots of the blade, each card is hosting two GPUs. The Tesla M6 GPU accelerator requires the NVIDIA GRID software license to function. This software license requirement is new for the M6 and lets you virtualize most applications, including professional graphics applications as required in this environment.

Figure 9-4 shows the C7000 enclosure and related components.

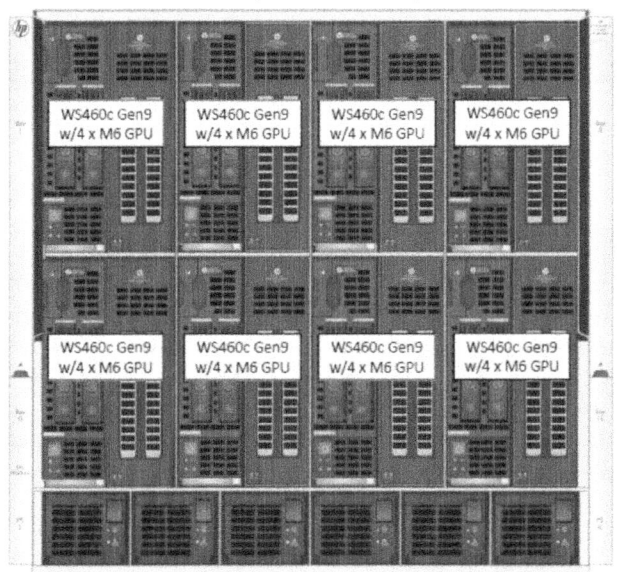

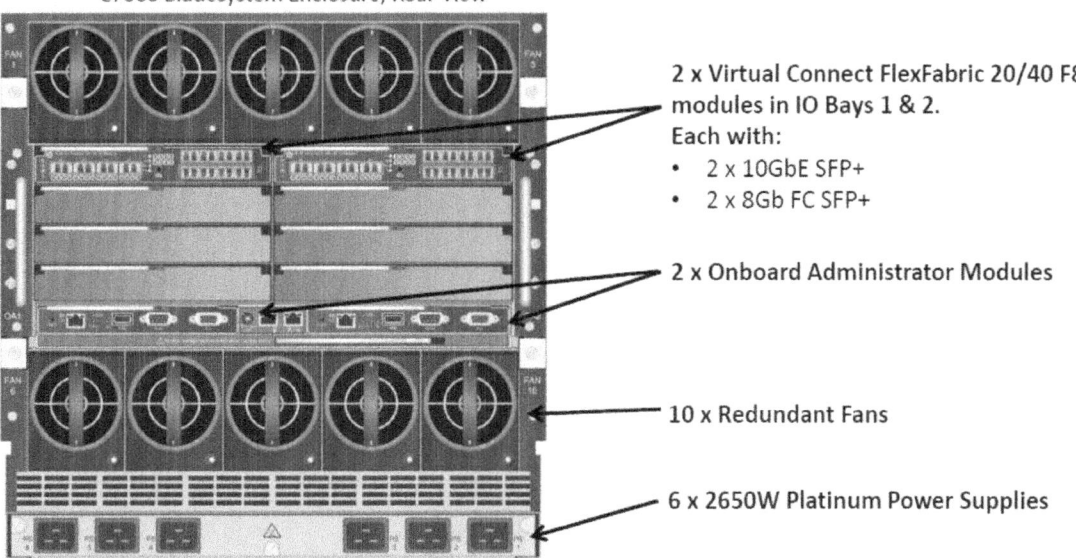

WS460c Gen9 blades with Graphics Expansion are virtualization hosts and have the following attributes:
- 2 x E5-2667v3 8C 3.2GHz CPUs
- 384GB 2133 MHz DDR4 memory (12x32GB LR DIMMs)
- 2 x 20Gb converged network ports
- 2 x 120GB SSD drives
- 4 x Multi GPU Maxwell based NVIDIA M6 graphics cards

2 x Virtual Connect FlexFabric 20/40 F8 modules in IO Bays 1 & 2. Each with:
- 2 x 10GbE SFP+
- 2 x 8Gb FC SFP+

2 x Onboard Administrator Modules

10 x Redundant Fans

6 x 2650W Platinum Power Supplies

Figure 9-4 Individual WS460-Gen9 Building Block

Datacenter considerations

Planning for power and cooling of a densely configured environment is critical to the long-term operation of the solution. Although racks, power, and cooling were already in place in the data center, it was important that the data center be equipped with adequate power and cooling for each rack. Each BLc7000 enclosure filled with 8 WS460-Gen9s required the following power and cooling.

Table 9-1 Power and cooling requirements (C7000 enclosure with 8—WS460-Gen9 blades)

System VA Rating	System BTU HR	Input system current	100% Utilization power	Idle input power	Max load input power
7036.9VA	23515.9 BTU	33.83 A	6896.16 W	1662.54 W	9535.63

HPE proposed a tightly integrated solution combining the HPE Workstation blades with software to provide the complete eVDI solution. The following are the key software components for this eVDI solution:

Citrix XenDesktop—includes Citrix HDX technologies that provide optimal performance while minimizing network bandwidth and latency. It also has capabilities such as HD video and 3D graphics optimization, universal printing, and profile management, with built-in WAN optimization to deliver a native-like experience to users on any device over any network connection.

HPE hardware optimized for Citrix XenDesktop—The HPE solution includes servers and storage that are specially tuned for running Citrix XenApp and XenDesktop. This solution includes:

- Blade infrastructure (described earlier).
- Storage (existing HPE 3Par StorServ 8450 4N all flash array).
- Citrix NetScaler was used to provide: RSA Two Factor Authentication, Traffic Prioritization, optimization and SSL.
- Symantec security solutions, including antivirus and spyware, Proactive, and Network thread protection, were already in use by this customer and also utilized in this solution.
- Liquidware Labs Stratusphere™ UX is used to monitor the desktop performance and user experience. This is essential to proactively addressing issues that could impact the desktop user, complementing Citrix XenApp.

In addition, several software components were in place including Quest Virtual Directory Server, Active Role Server, and group policies. VMware Horizon provides additional controls and mechanisms to prevent the transfer of data between the endpoint and the data center, keeping applications and sensitive information secured.

How did we arrive at this solution?

This solution was crafted based on evaluation of existing workloads to refine the configuration requirements. User demographics were provided to help with the definition of the workload and user segmentation for the eVDI workstation. This data included user counts, disk storage requirements, CPU requirements, and graphics, broken down into categories of High Graphics, High Compute, and High Compute and Graphics. Specific network bandwidth requirements were determined to be redundant 20 GbE Ethernet connections between Virtual Connect FlexFabric 20/40 F8 modules within each BLc7000 enclosure and the Top of Rack (ToR) switches. In addition, SAN connectivity would require redundant 16 GB FC connections for each enclosure to the Brocade-based FC SAN fabric.

The 3Par Storage solution was designed to allow for the required IOPs and capacity of the proposed environment and the ability to scale the environment with additional drives as needed. Additional trays and controllers may be added as the environment increases in size. This storage solution is based on a Solid State (SSD) drive footprint that will help to accelerate the IOPs in the environment.

To support the total required IOPs for the initial workstation environment, a 3Par StoreServ 8450 4-node solution was configured, along with 48 x 480 GB SSD drives.

With 48 total drives, configured in Raid 5 sets of eight drives, the total performance of the array exceeds 125,000 IOPs with a latency below .8 ms, providing more than enough performance for the currently defined workload. As additional workstations are added, additional enclosures and drives can be added up to 120 drives while maintaining the same performance.

In addition, 2-24 Port Brocade switches were provided to support the 3Par connections to the BLc7000 enclosures.

Each configured individual workstation blade provides redundant 8 GB SAN connections to the enclosure along with redundant 10 GB network connections supporting the VLAN traffic for both the graphics and management (Figure 9-5).

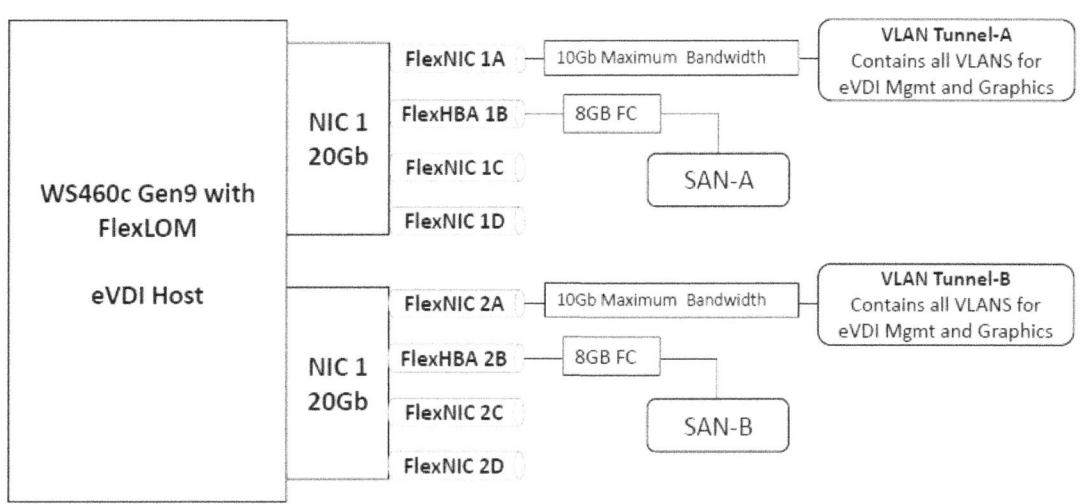

Figure 9-5 Network and SAN Connectivity

The design was crafted and tested as a combination of performance, reporting of the existing workstation environment along with benchmarks as part of a Proof-of-Concept. Based on the testing, we determined that four vCPUs, 24 GB of memory and 100 GB of disk were required per user. Using the WS460-Gen9, outfitted with 16 vCPU and four GPUs, we are able to support four eVDI users per physical workstation. Although faster processors and a higher memory density were available in this blade, the limit of four GPUs was what limited each workstation to four eVDI users. Table 9-2 summarizes this information.

Table 9-2 Customer requirements

Customer requirements	Detail
Users	100+
RAM per desktop	24 GB
vCPU per desktop	4 vCPU 3.2GHz
Storage	200GB vHD (100 IOPs)
Desktop Image Size	<=100 GB
Graphics (GPU Acceleration)	4GB vGPU delivered via NVIDIA Tesla M6

Figure 9-6 depicts the components of which eVDI solution is comprised.

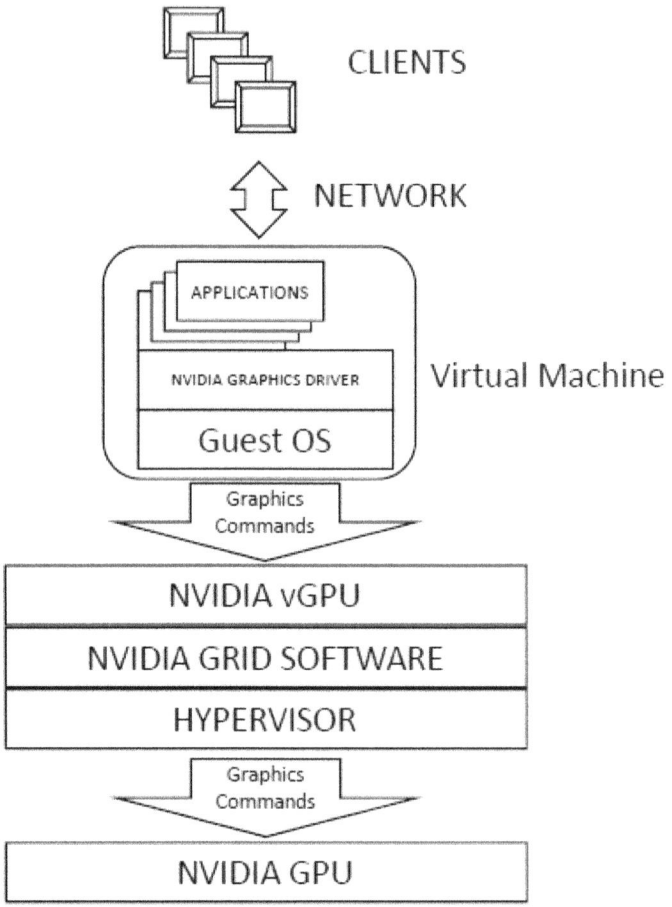

Figure 9-6 eVDI Component Stack with NVIDIA GPU integration

Processor selection

Selecting the correct processor for this eVDI solution required selecting the fastest processor available while providing a core count that allowed us to maximize the GPU utilization. In this case, two M6 GPUs supported a user to workstation ratio of 4:1 per workstation.

As newer processors and GPUs emerge in the industry, a higher ratio may be possible in the future.

Disaster Recovery (DR) considerations

Although at the time of publishing this solution, the customer had not yet completed their design for DR, the following were the options under consideration for the purpose of the DR planning:

- Provide 100% duplicated resources at the DR site.
- Provide a percentage of resources at the DR site allowing a first-come-first-served approach to the Virtual Workstations.
- Provide a subset of resources to support a predetermined list of "VIP" users.

Solution inventory

Table 9-3 shows the Bill of Material (BOM) for this eVDI application for a 16-node *building block*, providing high availability by splitting each building block across separate BLc7000 enclosures. As the solution is scaled up beyond 16 nodes, additional enclosures can be configured to maintain this division of resources. A 16-node building block is shown in Table 9-3.

Table 9-3 Bill of Materials

Blade Infrastructure		
Qnty	Part number	Description
2	681844-B21	HP BLc7000 CTO 3 IN LCD Plat Enclosure
2	E5Y41A	HPE OneView 3yr 24x7 Enclosure 16 Server License
4	691367-B21	HP BLc VC FlexFabric-20/40 F8 Module
8	AJ718A	HPE 8Gb Short Wave FC SFP+ 1 Pack
8	455883-B21	HP BLc 10G SFP+ SR Transceiver
2	733460-B21	HPE 6X 2650W Platinum Hot Plug Power Supply Kit
2	456204-B21	HP BLc7000 DDR2 Enclosure Management Option
2	677595-B21	HP BLc 1PH Intelligent Power Module Option
2	517520-B21	HP BLc 6X Active Cool 200 Fan Option
1	HA113A1	HPE Installation Service
2	HA113A1 5FY	HP Startup BladeSystem c7000 Infrastructure Service

Table 9-3 Continued.

Workstation Blades

Qnty	Part Number	Description
16	836738-B21	HPE WS460c Gen9 E5-v4 CTO Expansion Blade
16	819850-L21	HPE BL460c Gen9 E5 -2667v4 FIO Kit
16	819850-B21	HPE BL460c Gen9 E5-2667v4 Kit
192	805351-B21	HPE 32GB 2Rx4 PC4-2400T-R Kit
32	816965-B21	HP 120GB 6Gb SATA 2.5 MU-PLP SC S2 SSD
16	700764-B21	HP FlexFabric 20Gb 2P 650FLB Adapter
16	761871-B21	HP Smart Array P244br/1G Controller
16	775168-B21	HP Gen9 Expansion Blade Slot2 Enablement Kit
32	805133-B21	HP MultiGPU Carrier 2 Tesla M6 Adapter

Software components

The software stack for eVDI is just as important as the hardware components. Table 9-4 shows the key software components but is not comprehensive with all software.

Table 9-4 List of the software used to implement this eVDI environment

Item #	Product description	Additional details
1	Citrix: XenApp	Application virtualization
2	Citrix XenDesktop	Desktop virtualization
3	Citrix NetScaler	Application load balancer
4	VMware	ESXi 6.0 Virtualization OS
5	Liquidware Labs Stratusphere™ UX	Reporting and Monitoring

Many of these software components were described earlier with respect to their eVDI releveance.

Summary

This eVDI solution, as it compares to the HDI and general VDI covered in other chapters, is a perfect example of each specific workload having an optimal solution. In this case, the Workstation Blades were the ideal solution along with the software stack listed. It may be that your environment needs a combination of eVDI and general VDI in which case the ideal platforms on which each of these run would be deployed.

10 Citrix Implementation on HPE Hyper Converged 250

INTRODUCTION

This chapter covers the design and implementation of a Citrix XenApp 7.6 environment with multi-site access. The objective of the environment is to bring two completely separate healthcare facilities into a unified remote access infrastructure with Disaster Recovery (DR).

To provide the highest level of redundancy and availability, existing dark fiber between a primary site and a secondary site was used. The design takes advantage of clustering capabilities available in the HPE Hyper Converged 250 System (HC-250 in this chapter), allowing a VMware cluster to stretch between two locations. In this case, Site 1 and Site 2 are connected via Network RAID. This allows the two locations to be treated as a single site and enables VMware features such as High Availability (HA) and fault tolerance to be implemented between both sites thereby providing fully automatic and transparent DR functionality between these two locations. Introducing the capabilities of the stretched cluster, the two sites could either be configured to operate as active/active or active in the primary site and standby in the secondary site.

The following are some key benefits that this HC-250-based solution provides and the reason that it was selected for this solution:

- Capability of providing an appliance with a fixed amount of compute and the capability of scaling in a preconfigured consumption model.
- Reducing the amount of infrastructure required in order to achieve the end result; therefore, reducing the amount of data center devices and associated cost of such parameters as floor space, cooling, power, and management activities.
- The HC-250 is delivered preconfigured and has the capability of running multiple workloads on Day One.
- Uses the solution's capability of providing a storage stretched cluster in conjunction with VMware Metro Cluster.

CHAPTER 10
Citrix Implementation on HPE Hyper Converged 250

Solution overview

The HC-250 configuration used for this solution is shown in Table 10-1.

Table 10-1 HC-250 High-level components

	All Flash Hyper Converged 250
Rack Footprint	2U per appliance
Number of nodes per appliance	4 compute/storage nodes
Number of drives per node	6 Drives
Raw Capacity per node	9.6TB All-Flash (6x1.6TB SSD)
Usable Capacity by Node	7.13TiB
Usable Capacity by Cluster R10	14.16TB
Cluster CPUs	96 cores @ 2.5GHz
Cluster Memory	2TB DDR4 Memory (4x512GB)
Storage Controller Cache	4x 4GB FBWC
Network Ports (10GbE)	8x SFP+ ports
Network Ports (1GbE)	8x RJ-45 (1000BASE-T)
iLO ports	4x RJ-45 (100BASE-T)
Power supplies	2x 1400W Platinum Plus Power Supplies (High-line AC only, 240V)

These components fit into a rack as depicted in Figure 10-1.

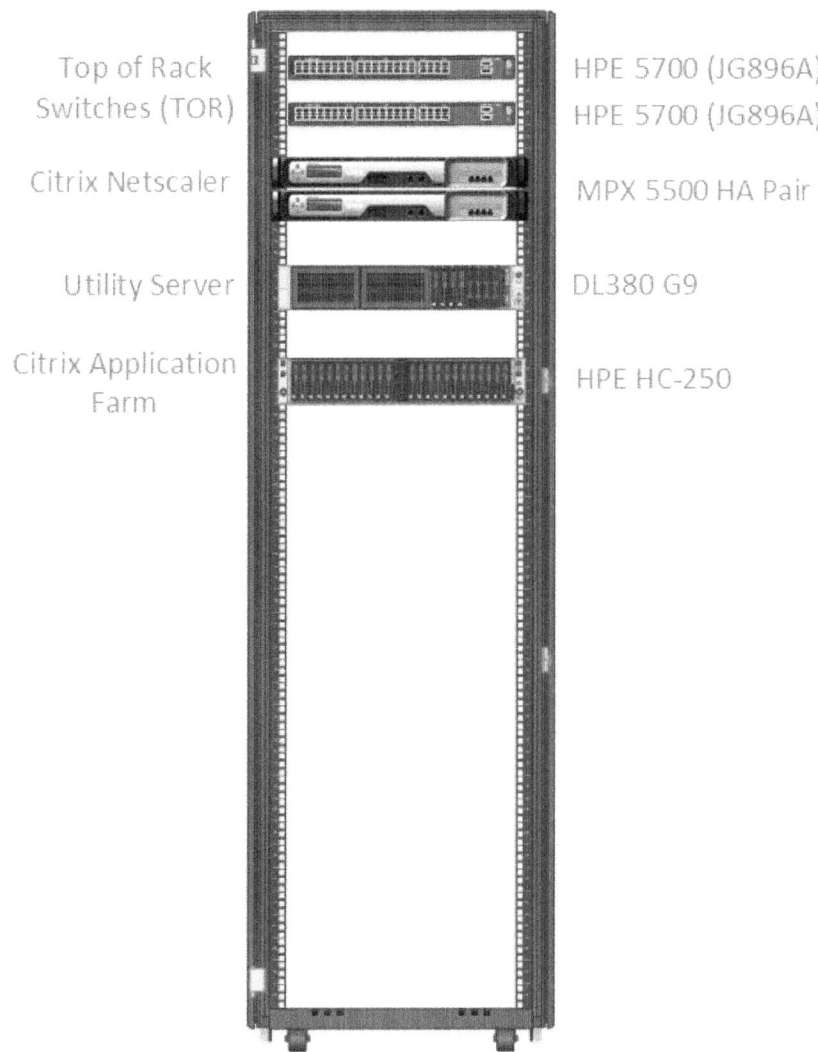

Figure 10-1 Rack diagram depicting a single site

Design for Multiple locations

The Citrix XenApp 7.6 multi-location configuration uses NetScaler MPX 5550 in HA pairs for remote access. One XenApp 7.6 site is created that spans across the two datacenters. This is achieved by connecting the two data centers and configuring Active Directory (AD) trusts. StoreFront is used to aggregate resources from any of the data centers and to provide users with a single point of access through the NetScalers. Citrix NetScaler accelerates application performance, load balances servers, increases security, and optimizes the user experience.

CHAPTER 10
Citrix Implementation on HPE Hyper Converged 250

In Figure 10-2, for each data center, there will be two NetScalers that are used to provide a high-availability configuration. The NetScalers are configured for Global Server Load Balancing and positioned in the DMZ to provide a multisite, fault-tolerant solution. Each HA Pair will have an additional URL outside of GSLB for direct Management access to a particular data center externally.

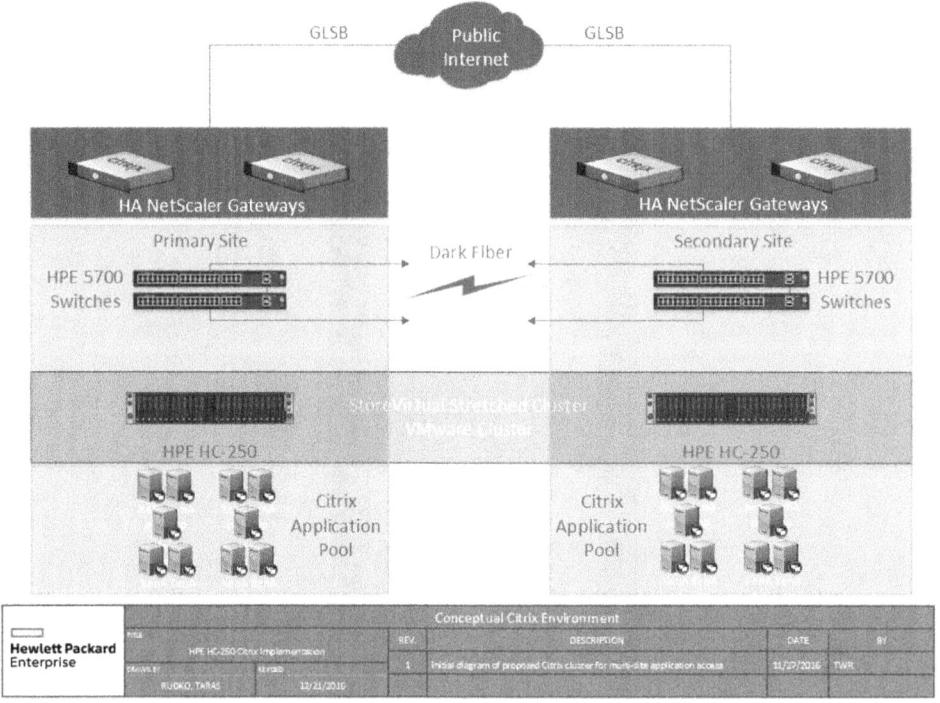

Figure 10-2 Conceptual diagram of two data centers

The next section describes some of the background related to crafting the design.

How did we arrive at this solution?

Several alternatives to the HC-250 could have been used for this design including C7000 BladeSystem and 3PAR, which is in place running other applications in this environment. The HC-250, however, was an ideal solution because all hardware and software components are preinstalled and preintegrated, and quick customization is achieved using the HPE OneView InstantOn software. After the initial installation, IT administrators managed the VMware vSphere environment on HC-250 with VMware vCenter, with which they are familiar, and HPE OneView for vCenter management integration.

HPE StoreVirtual technology

HPE StoreVirtual technology is built into the HPE HC-250 to achieve the desired HA in this solution. These considerations are summarized in the following list:

- The HC-250 has a compact form-factor that combines compute and storage in 2U.
- Familiar VMware vSphere management with VMware vCenter and HPE OneView for vCenter.
- HA provided by HPE StoreVirtual technology.
- All-flash is used to provide fast data access.
- 10 GbE networking for application and storage performance.
- First and only vSphere Metro Storage Cluster (vMSC) used between locations.

Figure 10-3 depicts a StoreVirtual stretched storage cluster implemented with the HC-250. Using network RAID, the storage nodes are able to stripe parity between the nodes as well as across the two physical sites. This feature enables all the storage to be presented in a single unified pool to all of the VMware ESXi nodes. The cluster is managed by a single vSphere vCenter instance and performs tasks between nodes. As an example, vMotion is performed across two sites. StoreVirtual and VMware are the base software components that support the implementation of the Citrix XenApp Farm using Virtual Machines (VMs). This is depicted in Figure 10-3.

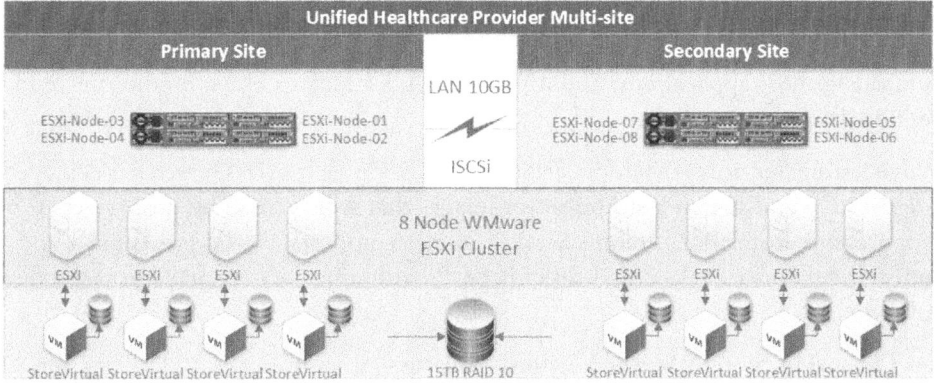

Figure 10-3 Solution overview

Citrix XenApp Environment

The key components used in the Citrix Farm on HC-250 are itemized below. This design is based on recommended best practices and past experience as shown in the following overview:

Sites: Two sites (primary and secondary) are configured in Active/Active with failover to each other. All users will connect to a single URL where applications will be presented in Storefront based on AD group permissions. If the primary site fails, the users will be launching their applications from the secondary site. If the secondary site goes down, then users will be launching their applications from the primary site.

Citrix Environment—One XenApp 7.6 Server environment with infrastructure servers in each site, two NetScaler HA pairs at each site.

In the primary site, there will be one HA pair. This provides local high availability, load balancing, and global server load balancing between both data centers during normal business. In the event of a failure at the primary site, users will fail over to the secondary site. In the event of a secondary site failure, users will failover to the primary site.

Citrix Storefront—Multiple Citrix Storefront servers will be deployed in each datacenter for local high availability, and enumeration of published applications from the XenApp 7.6 environment from each site with resource aggregation configured.

Citrix Desktop Delivery Controller: Multiple Delivery Controller servers will be deployed in each data center. The Delivery Controller consists of services that communicate with the hypervisor to distribute applications and desktops, authenticate and manage user access, and broker connections between users and their virtual desktops and applications. The Controller manages the state of the application delivery servers, starting and stopping them based on demand and administrative configuration

Citrix Studio—Multiple Citrix Studio management consoles will be deployed in each datacenter. This is the management console that enables configuration and management of the XenApp 7.6 farm. Studio provides wizards to guide you through the process of setting your environment, creating your workloads to host applications, and assigning applications to users. Studio can reside on the Citrix desktop delivery controllers.

Citrix Provisioning Services—Multiple dedicated Provisioning Services servers are deployed in each data center to store and stream XenApp server images that will be used by host servers for users to launch applications from. Provisioning Services allows computers to be provisioned and reprovisioned in real time from a single shared-disk image. Provisioning Services manages target devices as a device collection.

Windows Operating System Server Version—This design confirms that all Citrix infrastructure servers be built on Windows Server 2008 R2, Standard or Datacenter Editions, based on on-site discussions.

Citrix Data Store—The databases required for this design are the following: Site, Configuration Logging, Monitoring Logging, and Provisioning Services databases. These databases are hosted on a dedicated Microsoft SQL 2012 server with AlwaysOn Availability groups, this supports local HA. The benefits of AlwaysOn Availability Groups are detailed in the following Microsoft article: http://msdn.microsoft.com/en-us/library/hh510230.aspx. If it is determined that Microsoft SQL 2012 server is not available within the environment than Microsoft SQL 2010 r2 will be deployed in a mirrored setup.

Citrix License Server—A Citrix License server needs to be accessible from each site

Microsoft Remote Desktop Services (RDS) License Server—RDS License server is deployed in each data center on the same server used by the Citrix License Server.

Citrix NetScaler Gateway—NetScalers host the internal and external virtual IP address (using SSL) for users to login into, and then pass the users credentials (two factor authentication) to the Citrix Storefront servers to allow for application enumeration based on AD group membership.

Single-Sign On—Imprivata OneSign server and agent software will be configured to provide: strong authentication, single sign on for a single point of authentication management, password policy, and compliance reporting for all applications requiring authentication deployed within the XenApp environment.

Two Factor Authentication—RSA Adaptive Authentication Adapter for Citrix NetScaler will be configured and integrated to provide a comprehensive risk-based authentication and intrusion protection platform that balances security, usability, and cost. Adaptive Authentication monitors and authenticates users activities based on risk levels, institutional policies, and customer segmentation.

Citrix Director—Multiple Citrix Director Servers are deployed to enable IT support and help desk teams to monitor the XenApp environment, troubleshoot issues before they become system-critical and perform support tasks for end users.

Citrix Virtual Delivery Agent (VDA)—Installed on server operating systems, the VDA enables connections for applications. Please note in this release of XenApp 7.6, the VDA is only supported on the following Windows Server Operating System Versions: Windows Server 2012 R2 (Standard and Datacenter Editions), Windows Server 2012 (Standard and Datacenter Editions), Windows 2008 R2 SP1 (Standard, Enterprise, and Datacenter Editions).

Virtualization Resources—The following virtualization platforms are supported by XenApp 7.6: XenServer version (6.2, 6.1, and 6.0.2), VMware vSphere version (5.5, 5.1 Update 2, and 5.0 Update 2), Microsoft System Center Virtual Machine Manager version (2012 R2, 2012 SP1, and 2012), Cloud environments supported are Amazon Web Servers and Citrix CloudPlatform.

Hardware and software requirements

Table 10-2 shows the BOM for this design. This BOM includes the hardware and some of the software for this solution. Some of the software was procured outside of the HPE BOM.

Hardware requirements

The following minimum hardware is recommended for each of the locations. Network capability of 10 GB is already operational at each of the three data centers:

o HC-250 Appliance with redundant power supplies

- 2 x HPE 5700 40XG switches
- Each node configured with (2) 10 GB and (2) 1 GB Ethernet ports
- (1) ILO port per node for out-of-band management

CHAPTER 10
Citrix Implementation on HPE Hyper Converged 250

Table 10-2 BOM for design

Qty	Part No	Description
1	H6J66A	HPE 42U 600x1075mm Advanced Shock Rack
1	H6J66A 001	HP Factory Express Base Racking Service
1	M0T03B	HPE HC 250 System for VMware vSphere Sys
1	M0T03B 0D1	Factory integrated
4	M0T04B	HPE Hyper Converged 250 Node
4	M0T04B 0D1	Factory integrated
4	793028-B21	HP XL1x0r Gen9 E5–2680v3 Kit
4	793028-B21 0D1	Factory integrated
4	793028-L21	HP XL1x0r Gen9 E5–2680v3 FIO Kit
64	728629-B21	HP 32GB 2Rx4 PC4–2133P-R Kit
64	728629-B21 0D1	Factory integrated
24	804631-B21	HP 1.6TB 6G SATA MU-2 SFF SC SSD
24	804631-B21 0D1	Factory integrated
4	665243-B21	HPE Ethernet 10Gb 2P 560FLR-SFP+ Adptr
4	665243-B21 0D1	Factory integrated
4	P9B51A	HPE HC 250 SW LTU for VMware vSphere 6.0
4	P9B51A 0D1	Factory integrated
1	676277-B21	HP 36pin Suv Dongle Cord Kit
1	676277-B21 0D1	Factory integrated
2	720620-B21	HPE 1400W FS Plat Pl Ht Plg PS Kit
2	720620-B21 0D1	Factory integrated
1	H1K92A3	HPE 3Y Proactive Care 24x7 Service
1	H1K92A3 YMW	HPE CS 250-HC StoreVirtual System Supp
4	H1K92A3 YMX	HPE CS 250-HC StoreVirtual Node Support
4	H1K92A3 YMY	HPE CS 250-HC StoreVirtual SW LTU Supp
1	HA114A1	HP Installation and Startup Service
1	HA114A1 5WG	HPE 200 s HC StoreVirtual Startup SVC
1	H6J85A	HPE Rack Hardware Kit
1	H6J85A 0D1	Factory integrated
2	H5M58A	HPE Basic 4.9kVA/L6–30P/C13/NA/J PDU
2	H5M58A 0D1	Factory integrated

Table 10-2 Continued

Qty	Part No	Description
1	BW932A	HPE 600mm Rack Stabilizer Kit
1	BW932A B01	Include with complete system
1	BW930A	HPE Air Flow Optimization Kit
1	BW930A B01	Include with complete system
1	BW906A	HPE 42U 1075mm Side Panel Kit
1	BW906A 0D1	Factory integrated
1	120672-B21	HPE Rack Ballast Kit
1	120672-B21 0D1	Factory integrated
1	HA113A1	HPE Installation Service
1	HA113A1 5BY	HPE Rack and Rack Options Install SVC
1	P8A90AAE	HPE SV VSA 2014 4TB 3pk 3yr DPS E-LTU
1	P9S43AAE	HPE RMC for HC Systems E-RTU
10	P9T49AAE	VMw Horizon Std 10pk 3yr CU E-LTU
1	H1K92A3	HPE 3Y Proactive Care 24x7 Service
10	H1K92A3 RXZ	HPE VMw Horizon View 10Pk 3yr ESW Supp
1	H1K92A3 YN9	HPE SV VSA 2014 4TB 3pk 3yr LTU Support
10	JD097C	HPE X240 10G SFP+ SFP+ 3m DAC Cable
2	JH061A	HPE 57xx CTO Switch Solution
2	JG896A	HPE 5700 40XG 2QSFP+ Switch
2	JG896A 0D1	Factory integrated
4	JG900A	HPE A58x0AF 300W AC Power Supply
4	JG900A 0D1	Factory integrated
4	JG900A ABA	U.S. - English localization
4	JD092B	HPE X130 10G SFP+ LC SR Transceiver
4	JD092B 0D1	Factory integrated
2	JG326A	HPE X240 40G QSFP+ QSFP+ 1m DAC Cable
2	JG326A B01	Include with complete system
4	JC682A	HPE 58x0AF Bck(pwr) Frt(prt) Fan Tray
4	JC682A 0D1	Factory integrated
1	H1K92A3	HPE 3Y Proactive Care 24x7 Service
2	H1K92A3 TF4	HPE FF 5700 Switch Support
1	HA114A1	HP Installation and Startup Service
2	HA114A1 5RN	HPE Top of Rack Startup SVC

- Four compute nodes each being a VMware Host.
  - 2 physical sockets
  - 24-cores, 2.5 Ghz Procs
  - 512 GB of Ram
  - Access to Necessary Networks

Management software requirements

The following software must be hosted on a production environment that has access to the physical servers. It cannot be hosted on any of the hosts that it will be managing.

- vCenter
- OneView

Microsoft licensing requirements

To simplify management and reduce the need to implement future upgrades, Windows 2012 will be deployed on all management and support servers while Windows 2008 R2 is deployed on the application publishing servers. This ensures the future-proofing of this infrastructure while leveraging the significant investment made in Windows 2008 TS licenses.

General requirements

- Number of Physical Hosts is estimated based upon number of user sessions to be supported.
- Access to necessary networks.
- Spread out Infrastructure Servers among the Hosts.
- vSphere for Desktops licenses to be used.
- Dedicated vCenter to be hosted in Site 1 with HA failover to Site 2 in production cluster.
- OneView VM needs is hosted in production cluster.

Summary

This solution is a perfect example of how a Hyper Converged system with software such as VMware and Citrix can be rapidly deployed to get two sites up-and-running quickly. Disaster Recovery and High Availability were key design considerations to ensure that user productivity would not be interrupted in the event of a site failure. The implementation of the HC-250 is highly streamlined because of all the hardware and software components that can be preinstalled, preconfigured, and tested before they arrive on site.

11 EMR Software on HPE ConvergedSystem

INTRODUCTION

The most basic function of Electronic Medical Records (EMR) Software is to provide a platform to healthcare providers that will manage and share patient records. Most EMR solutions do far more than simple records management, but this chapter will focus on design considerations involved with a specific software platform rather than the detailed capabilities of that software. In this chapter, Epic is the EMR software, which is targeted toward medium-to-large healthcare practices and hospitals.

The hardware configurations for Epic vary based on requirements and software modules that will be deployed. A common Epic environment is heterogeneous and consists of dozens of virtual and/or non-virtualized servers connected to a high-speed Fibre Channel SAN and shared storage device.

ConvergedSystem 700

This Epic solution is based on the HPE ConvergedSystem 700 (CS700). HPE ConvergedSystem 700 ships with factory-integrated server, storage, networking, and management components—all preconfigured to deliver speed and simplicity. The system also includes integrated lifecycle management compliance, all the way to the workload level, as well as solution-level support with single-vendor accountability.

The HPE CS700 used in this environment has the following characteristics:

- Rapid deployment
    - A factory-configured ConvergedSystem, integrating server, storage, networking, and management components—all built and tested together before being shipped to the customer site.
    - The final configuration included purpose-built elements to deliver faster time-to-value, reducing deployment times from months to weeks. This helped ensure that virtualization, cloud, or workplace productivity services were brought online faster.

>  **Note**
> Non-converged systems must be shipped as individual components, setup and connected on-site, which would have greatly extended the deployment timelines. A rapid, secure, and modular approach was required to gain greater control of patient data and ensure it would get into the right hands, at the right place, and the right time.

- Simplified system management
    - A single point of management for server and storage devices to simplify day-to-day management tasks as well as end-to-end component compatibility.
    - A single management console to simplify management tasks across administrators, equipment, and processes, as well as integrating with VMware vCenter to create.
- High performance
    - Shared storage that can scale to meet performance and capacity requirements without the need to add arrays.
    - High IOPs that are beneficial to the performance of Epic.
- Flexibility
    - HPE has a wide selection of server options to fit the requirements for a wide range of compute workloads.
    - Epic requires certain workload isolation and thus needs to use separate pools of storage. That capability, present here, is missing in many flash arrays.
- Floor and rack space consolidation
    - BladeSystems increase server density per rack unit and decreases connectivity complexity.
    - All Flash Arrays provide more capacity and performance in a smaller footprint while lowering power and cooling costs.
- Validated component compatibility
    - ConvergedSystem provide certified hardware, firmware, driver, and operating system compatibility.

Figure 11-1 depicts this solution at a high-level.

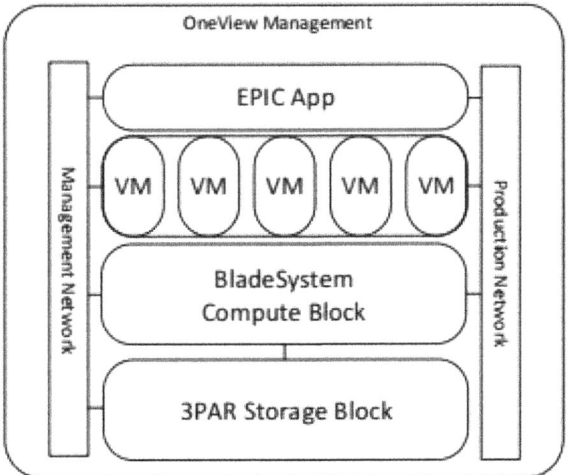

Figure 11-1 High-level EMR block diagram

Solution overview

The following are the CS700 components used in this solution:

HPE ConvergedSystem 2.0 solution with:

- 3 x HPE BladeSystem c7000 Enclosures
- 27 x HPE ProLiant BL460c Gen9 Server Blades each with:
    - 2 x Intel Xeon E5-2697v3 2.6 Ghz 14 core CPUs (per Epic Hardware Configuration Guide)
    - 256 GB memory using 8 x 32 GB dual in-line memory modules (DIMM)
    - 1 x HPE FlexFabric 20Gb 2P 650FLB FIO adapter
    - HPE Dual 8GB microSD EM Universal Serial Bus (USB) kit
- 2 x HPE ProLiant DL360 Gen9 Server Management nodes
- TOR switch solution consisting of
    - 2x 10/40 GbE switches (Production network)
    - 2 x 1 GbE switches (Management network)
- HPE 3PAR StoreServ 7440c 4N storage solution with:
    - 4 controller nodes
    - 8 x HPE Virtual Connect 8GB FC ports per node (32 total)
    - 6 x M610 SFF drive trays

CHAPTER 11
EMR Software on HPE ConvergedSystem

- o Two SSD disk pools
    - 32x HPE 920GB SED SSD for Epic Systems database (Epic Systems pool P0)
    - 32 x HPE 1.92 terabyte SED SSD for all other Epic usage (Epic Systems pools P1-P5)
    - HA for all CPGs
    - Manual disk filtering to isolate pools (Epic Systems requirement)
- o HPE 3PAR StoreServ software included:
    - Base Operating System Suite
    - System Reporter
    - Dynamic Optimization
    - Priority Optimization
    - Data Encryption
    - Replication Suite
    - Application Suite for VMware
    - Application Suite for Microsoft SQL
- vCenter plus VMare vSphere with Operations Management Enterprise Plus licenses included
- Additional management software included:
    - o HPE StoreFront Analytics for HPE 3PAR
    - o HPE Operations Analytics for HPE OneView
    - o HPE Virtualization Performance Viewer for HPE OneView
- Three rack configuration
- 3 Phase 30 Amp power distribution unit used on all racks
- 5 year ProActive Care 24x7 support with DMR on all above

These components are shown in the rack diagrams in Figures 11-2 and 11-3:

TRIED AND TESTED SOLUTIONS TO ACCELERATE IT AND BUSINESS TRANSFORMATION | 115
Ideal Platforms for Optimizing IT Workloads

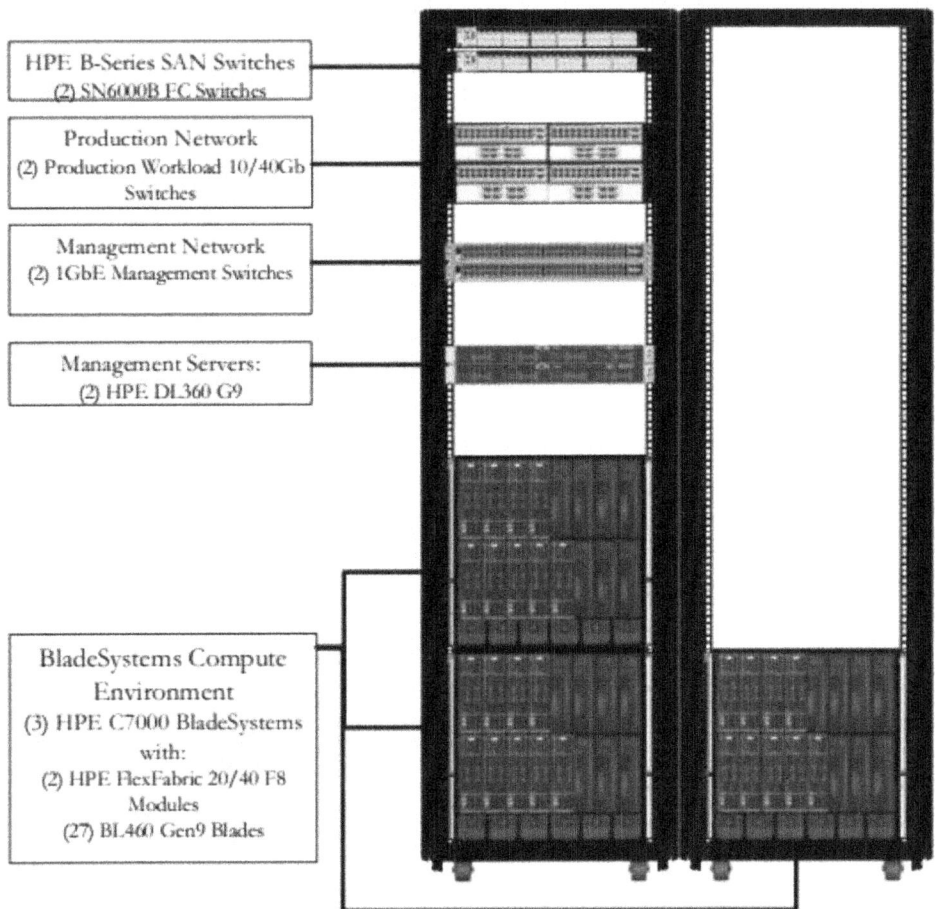

Figure 11-2 Rack diagram depicting the Epic compute environment

CHAPTER 11
EMR Software on HPE ConvergedSystem

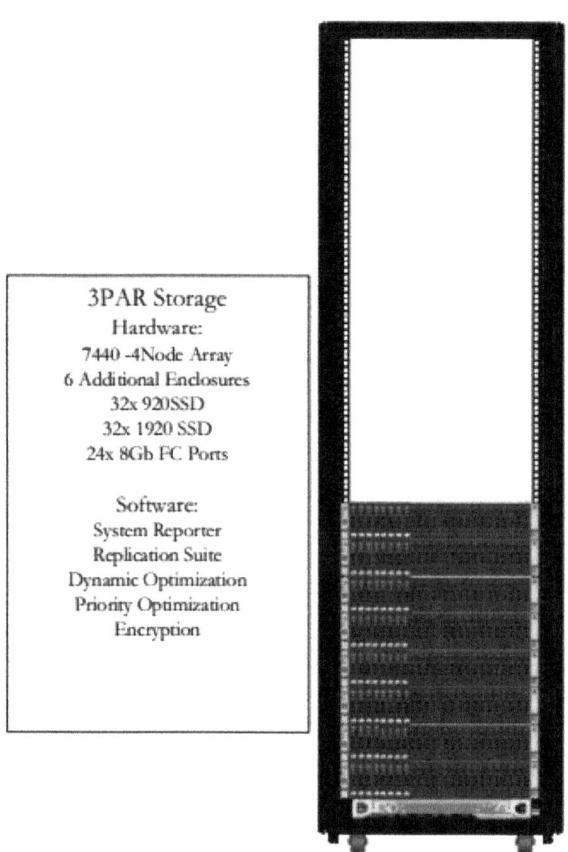

Figure 11-3 Rack diagram depicting the Epic storage environment

Design considerations

Epic is a performance-sensitive application that requires low latency from the storage array. The solution in this chapter is for a primary site deployment. Disaster Recovery and Backup components, while important factors in the design, will not be addressed in depth.

Server and storage requirements are defined in the Epic Hardware Configuration Guide that acts as the basis for sizing in the design.

Compute platform design considerations

This deployment uses 27 blade servers across three HPE BladeSystem chassis that are in the initial deployment. Each rack server requires the following:

- Redundant FC connectivity
- Redundant production LAN connectivity

- Redundant management network
- Additional port for HPE's Integrated Lights Out (iLO) management solution

If rackmount servers had been used, the SAN connectivity would require a total of 54 ports for servers and additional 24 for the shared storage array for a total of 78 ports. BladeSystem provides SAN port consolidating with as few as four SAN uplinks per chassis and still provide appropriate throughput to handle any workloads even with a fabric failure. Because this solution uses three blade chassis, the total number of SAN host uplink drops from 54 to 12. The decrease in port count correlates to a decrease in SAN switch cost and cabling complexity.

LAN connectivity similarly benefits from using blades versus rack mount servers. Rack mount server would require:

- 54-10 GbE ports
- 54-1 GbE ports for node management
- 27-1 GbE ports for iLO

BladeSystem decrease cabling to the following:

- Either two-40 GbE or 8–10 GbE uplinks per chassis for converged production and management traffic
- One iLO port per chassis.
- Two On-board Administrator Ports

Because there are three blade chassis in this configuration, the total port count would decrease from a combined 135 Ethernet ports to 33 total ports if using 10 GbE uplinks for the blades. Using 40 GbE uplinks would decrease the port count even further to a total of six-40 GbE and nine-1 GbE.

BladeSystem is a "wire once" technology. The solution discussed in this chapter uses 27 production servers spread across three blade chassis. Each BladeSystem has a capacity for up to 16 of the chosen server model (BL460 Gen9 in this case), so expanding the environment from the existing 27 servers to the maximum of 48 (Three Blade Systems with sixteen blade servers per chassis) requires no additional LAN or SAN connectivity costs or efforts.

Storage design considerations

An HPE 3PAR 7440 with Fibre Channel is used as the shared storage device in the CS700. Epic tests, validates, and endorses storage arrays from multiple manufacturers for compatibility with their application. The best endorsement of an array is to be placed in the "High Comfort" category. That means the Epic engineering team has tested the array with their workloads, and those arrays have been used successfully in customer environments. HPE 3PAR arrays are placed in the "High Comfort" category for any size Epic environment.

Epic consists of multiple databases with varied workloads, but in sizing the storage, the primary considerations for this solution are the following:

- Operational Database Platform (ODB)
- Relational database management system (RBDMS) for reporting

The ODB and RBDMS database workloads are the primary focal point in Epic designs to ensure high I/O performance. In addition, there are other Epic modules and applications for which design considerations are made.

ODP workload

The ODP workload is generally small block random IO, but the application has very high IO bursts. These bursts happen every 80 sec, and it is common to see them generate between 100K and 200K IOPS at the array. The array must be able to handle the content of this write burst in cache and destage rapidly in order to avoid performance degradation.

RDBMS workload

RDBMS workload will vary by ODB size but can run for several hours with sequential throughput reaching several hundred MB/s with a mix of read and write request. Reporting will generally be a read intensive workload with sequential I/O reaching into the 1 GB/s range. As with most server-storage interdependencies, throughput can greatly increase or decrease depending on the ability of the server to process data and the storage arrays ability to handle request. The performance capabilities of the array decreased Extract Transform Load (ETL) and reporting times from hours to minutes in most cases.

Snapshots and clones

Epic uses LUN cloning for backup in order to offload backup I/O from the production disk pool. The performance provided by SSD allows for the use of snapshots instead of clones, decreasing backup windows and increasing recoverability in the case of data loss.

Deduplication

Deduplication is a common feature that can decrease the amount of raw storage required to store data. When looking into storage platforms, you always try to understand the efficiency of an array. Deduplication can play a part in many environment. Epic production and nonproduction environments will allow the customer to realize space savings with deduplication.

3PAR Four-Node 7440 AFA

The 7440 was chosen because it has all of the performance benefits of an all-flash array, but Hard Disk Drives (HDD) can be used in the array as well as SSD. 3PAR is one of the few all-flash-capable arrays that permit the isolation of a pool of SSD. In this solution, the all-flash 3PAR-backed ConvergedSystem was replacing old technology. A unique benefit to using a 3PAR is that the array, model dependent, can scale from two to four, six, or eight nodes. An array with four or more nodes allows the use of write cache even in the event of a node outage. Without caching, data, throughput can be negatively impacted and application users will notice the degraded performance. 3PAR can provide performance availability with the use of multiple node pair in a single array.

The 3PAR array is typically configured for one large pool of disk, but Epic mandates a separate storage pool for the ODP so the 3PAR is configured as follows:

- Four Node 7440 array
    - 32x920GB Solid State Drive (SDD) or ODP
    - 32x 1920GB SSD for other workloads
    - 24x 8Gb FC host port

What happens if your array cannot isolate and dedicate a pool of drives for a specific Epic workload? A second array is required. A second array running a production workload requires additional management, upgrade planning, and higher Extract, Transfer, Load (ETL) times, installation, and connectivity efforts. A single array with flexible configuration options is an optimal platform for Epic.

The unique cluster capabilities of 3PAR allow the array to scale by adding nodes and drives. The additional nodes are in an Active/Active configuration. That means you will have active paths from the host to the LUN through any and all controllers. This helps balance the workload automatically without assigning LUNs to active-failover controllers. Another benefit of the 3PAR cluster is cache coherency. The array cache is shared across more than two nodes, which is not possible with two node arrays. Cache Coherency is provided by a Full-Mesh controller backplane. This provides better performance in a healthy state and, more importantly, provides performance availability in case of a planned or unplanned outage. All storage platforms require some form of firmware or software upgrade as part of normal maintenance. Sometimes, those upgrades are to add features or to fix a discovered flaw, but all storage arrays require some form of software upgrade and that can cause node outages. Most of the arrays on the market are limited to two nodes. With that antiquated architecture, a node outage causes a path and controller failover that will increase the latency on the Epic workload. A four, six, or eight-node 3PAR will not experience that path and node failover due to the Active/Active.

Solid State Drives provide a massive increase in performance over HDD. Comparing standalone drives is not pertinent to this discussion, so we want to look at the difference in capacity and performance in a shared array. In this example, we will assume a workload of 50/50 Read/Write with random 8K request size. I will show the difference in efficiencies between the two drive types by differentiating RAID requirements and performance. Using a current four-node 3PAR, 32x 1.92TB SSD drives can provide nearly 350K Front End (delivered to the hosts) IOPS with the aforementioned workload. Without deduplication or and other compaction technologies, we will have over 43 TB of usable space out or 55.8 raw TB. Remember that 3PAR lets you choose your RAID set. We can use RAID 0, 1, 5, 6, so we can stretch usable capacity out to R5 (8+1) or a RAID 6 (14+2). That 43 TB usable fits in a 4U of rack space and includes RAID and Sparing overhead. Over One-Thousand 15K HDD would be required to match the workload capabilities of a few dozen SSD. This would require about four full racks and approximately 20 times the power of the all-flash array. At the time of this writing, the smallest 15K drives available are 300 GB 15K. In order to reach the performance of the SSD, the 15K drives would be in a RAID 1 set and about 270 TB raw would only yield about 130 TB usable. SSD is already compelling when looking at performance versus capacity but consider other use cases beyond Epic and compaction technologies that can provide many times the usable capacity when compared to spinning media.

Upon installation and testing of any new solution, Epic will run performance tests consisting of simulated Epic workloads on the server and storage to verify it meets application requirements. The testing showed that the HPE ConvergedSystem was able to deliver approximately six times the performance required by Epic, which leaves room for native growth without the need for replacing any platform with a more powerful solution. This is a primality due to the advantages of the four-node all flash 3PAR storage array.

Management and monitoring

HPE OneView is the glue that bonds the individual components together and transforms them into an appliance-like offering. OneView provides a RESTful API and provides integrated infrastructure management.

This is a VMware-based solution so we will look at the management integration for VMware environment. There are many other software components, such as operating system, that are not covered in this chapter. The following are key management components:

- HPE OneView provides the following:
  - Infrastructure reporting such as hardware inventory, server, storage profiles, alerts, and activities, firmware baselines
  - VMware VSOM including vRealize Operations Manager VMware Capacity Management and Performance Monitoring
- HPE Operations Analytics for HPE OneView provides the following:
  - Real-time troubleshooting of converged infrastructure, infrastructure capacity, and so on
  - Provides capabilities from triage and diagnosis to stakeholder analysis to decision support
- 3PAR System Reporter
  - Real-time and periodic long-term performance reporting of storage

- HPE StoreFront Analytics for 3PAR
  - Plug in to VMware vSphere VSOM for deep insight into storage
- HPE Virtualization Performance Viewer for HPE OneView for a unified view of the physical enclosure and virtual cluster layout
  - Assess risk to your cluster deployments
  - Reclaim waste and lower cost with optimization analysis
  - Manage and plan for capacity with forecast and what-if modeling

Implementation and support

One of the key objectives of this installation was to have an integrated system, in this case HPE ConvergedSystem that was delivered fully operational. This reduced the overall implementation dramatically which, in the case of a healthcare provider, is important so that the time to deploy EMR is reduced. Although this was a complex solution with many components in the BOM, the integrated ConvergedSystem and work done at the factory made this implementation go smoothly.

Combined hardware and software inventory

Table 11-1 is the combined hardware and software BOM:

Table 11-1 BOM for CS700

Quantity	Part number	Description
1	BW908A	HPE 42U 600x1200mm Enterprise Shock Rack
1	K9T75A	HP CS700 Virt 2.0 vw Kit.
1	H8A03A1-305	HPE CS700 2.0 Virt FE Deployment SVC
1	L2G47A	HP CS700 2.0 Virt Encl Kit
9	727021-B21	HP BL460c Gen9 10Gb/20Gb FLB CTO Blade
9	767049-L21	HP BL460c Gen9 E5-2697v3 FIO Kit
9	767049-B21	HP BL460c Gen9 E5-2697v3 Kit
72	726722-B21	HP 32GB 4Rx4 PC4-2133P-L Kit
9	700764-B21	HPE FlexFabric 20Gb 2P 650FLB FIO Adptr
1	E5Y41A	HPE OV 3yr 24x7 Encl FIO Phys 16 Svr Lic
56	AJ716B	HPE 8Gb Short Wave B-Series SFP+ 1 Pack
1	L2G47A	HP CS700 2.0 Virt Encl Kit
9	727021-B21	HP BL460c Gen9 10Gb/20Gb FLB CTO Blade

Table 11-1 Continued.

Quantity	Part number	Description
9	767049-L21	HP BL460c Gen9 E5-2697v3 FIO Kit
9	767049-B21	HP BL460c Gen9 E5-2697v3 Kit
72	726722-B21	HP 32GB 4Rx4 PC4-2133P-L Kit
9	700764-B21	HPE FlexFabric 20Gb 2P 650FLB FIO Adptr
1	E5Y41A	HPE OV 3yr 24x7 Encl FIO Phys 16 Svr Lic
8	AJ716B	HPE 8Gb Short Wave B-Series SFP+ 1 Pack
1	TK821A	HP CS Rack Side Panel 1200mm Kit
1	TK816A	HP CS Rack Light Kit
1	TK815A	HP CS Rack Door Branding Kit
1	TK732A	HP 42U PDU MANAGEMENT-BRACKETS Cable
2	QK753B	HPE SN6000B 16Gb 48/24 FC Switch
24	QK724A	HPE B-series 16Gb SFP+SW XCVR
2		1GbE Management Switch
4	JG330A	HPE X240 QSFP+ 4x10G SFP+ 3m DAC Cable
2	JG331A	HPE X240 QSFP+ 4x10G SFP+ 5m DAC Cable
2		10GbE/40GbE Production Switches
4	JG330A	HPE X240 QSFP+ 4x10G SFP+ 3m DAC Cable
2	JG331A	HPE X240 QSFP+ 4x10G SFP+ 5m DAC Cable
2	L2G46A	HP CS700 2.0 Virt Mngmt Kit
2	P8B31A	HPE OV w/o iLO 3yr 24x7 FIO Phys 1 LTU
2	BD505A	HPE iLO Adv incl 3yr TSU 1-Svr Lic
2	755997-B21	MS WS12 R2 Std FIO Npi E/F/I/G/S SW
1	H1K93A5	HPE 5Y Proactive Care 24x7 wDMR Service
2	H1K93A5#R2M	HPE iLO Advanced Non Blade - 3yr Support
2	H1K93A5#S8X	HPE MS WS12 Standard OS+APP Support
2	H1K93A5#SVP	HPE One View w/o Ilo Support
2	H1K93A5#U4T	HPE CS 700 2.0 Virt Man Kit Support
4	H8B52A	HPE Mtrd Swtchd 8.6kVA/L15-30P/NA/J PDU
2	C7536A	HP Ethernet 14ft CAT5e RJ45 M/M Cable
3	C7535A	HP Ethernet 7ft CAT5e RJ45 M/M Cable
4	C7533A	HP Ethernet 4ft CAT5e RJ45 M/M Cable
1	BW930A	HPE Air Flow Optimization Kit
1	BW902A	HPE Rack Baying Kit
1	BW891A	HPE Rack Grounding Kit

Table 11-1 Continued.

Quantity	Part number	Description
8	AJ836A	HPE 5m Multi-mode OM3 LC/LC FC Cable
6	AJ835A	HPE 2m Multi-mode OM3 LC/LC FC Cable
2	AF522A	HPE Intelligent 8.6kVA/L15-30P/NA/J PDU
2	433718-B21	HP BLc7000 10K Rack Ship Brkt Opt Kit
1	BW908A	HPE 42U 600x1200mm Enterprise Shock Rack
1	BW908A#001	HP Factory Express Base Racking Service
1	H8A03A1-306	HPE CS700 2.0 Virt Add Comp Rack FE SVC
1	L2G47A	HP CS700 2.0 Virt Encl Kit
9	727021-B21	HP BL460c Gen9 10Gb/20Gb FLB CTO Blade
9	767049-L21	HP BL460c Gen9 E5-2697v3 FIO Kit
9	767049-B21	HP BL460c Gen9 E5-2697v3 Kit
72	726722-B21	HP 32GB 4Rx4 PC4-2133P-L Kit
9	700764-B21	HPE FlexFabric 20Gb 2P 650FLB FIO Adptr
1	E5Y41A	HPE OV 3yr 24x7 Encl FIO Phys 16 Svr Lic
8	AJ716B	HPE 8Gb Short Wave B-Series SFP+ 1 Pack
1	TK816A	HP CS Rack Light Kit
1	TK815A	HP CS Rack Door Branding Kit
1	TK732A	HP 42U PDU MANAGEMENT-BRACKETS Cable
4	H8B52A	HPE Mtrd Swtchd 8.6kVA/L15-30P/NA/J PDU
2	C7536A	HP Ethernet 14ft CAT5e RJ45 M/M Cable
3	C7535A	HP Ethernet 7ft CAT5e RJ45 M/M Cable
3	C7533A	HP Ethernet 4ft CAT5e RJ45 M/M Cable
1	BW930A	HPE Air Flow Optimization Kit
1	BW902A	HPE Rack Baying Kit
1	BW891A	HPE Rack Grounding Kit
8	AJ836A	HPE 5m Multi-mode OM3 LC/LC FC Cable
6	AJ835A	HPE 2m Multi-mode OM3 LC/LC FC Cable
1	433718-B21	HP BLc7000 10K Rack Ship Brkt Opt Kit
2	JG811AAE	HPE VSR1001 Virtual Services Rtr E-LTU
58	P9U15AAE	VMw vSOM EntPlus 1P 5yr E-LTU
1	P9U42AAE	VMw vCenter Server Std for vSph 5y E-LTU
1	H1K93A5	HPE 5Y Proactive Care 24x7 wDMR Service
2	H1K93A5#2C2	HPE Networks SW Group 120 License Supp
2	H1K93A5#QAM	HPE SN6000B 16Gb 48/24 FC Switch Support

CHAPTER 11
EMR Software on HPE ConvergedSystem

Table 11-1 Continued.

Quantity	Part number	Description
1	H1K93A5#R62	HPE VMw vCntr Srv Std 5yr SW Support
58	H1K93A5#RX4	HPE VMw vSOM EntPlus 1P 5yr ESW Support
3	H1K93A5#SVQ	HPE One View for blades Support
27	H1K93A5#TT8	HPE BL460c Gen9 Server Blade Support
3	H1K93A5#U4S	HPE CS 700 2.0 Virt Enc Kit Support
2	TC356A	HPE SN6000B SAN Switch 12-port Upg LTU
1	BW908A	HPE 42U 600x1200mm Enterprise Shock Rack
1	E7X84A	HPE 3PAR StoreServ 7440c 4N St Cent Base
4	QR486A	HPE 3PAR 7000 4-pt 8Gb/s FC Adapter
8	E7Y52A	HPE M6710 920GB SFF FE SSD
8	K0F27A	HPE M6710 1.92TB SFF FE SSD
1	BD374A	HP 3PAR 7440c OS Suite Base LTU
64	BD381A	HP 3PAR 7440c OS Suite Drive LTU
1	BD382A	HP 3PAR 7440c Replication Suite Base LTU
64	BD383A	HP 3PAR 7440c Replication Ste Drive LTU
1	BD390A	HP 3PAR 7440c Dynamic Opt Base LTU
64	BD391A	HP 3PAR 7440c Dynamic Opt Drive LTU
1	BD402A	HP 3PAR 7440c Priority Opt Base LTU
64	BD403A	HP 3PAR 7440c Priority Opt Drive LTU
1	BD406A	HP 3PAR 7440c Data Encryption LTU
1	BD375A	HP 3PAR 7440c Reporting Suite LTU
1	BD376A	HP 3PAR 7440c App Suite VMware LTU
1	BD378A	HP 3PAR 7440c App Suite SQL LTU
6	QR490A	HPE M6710 2.5in 2U SAS Drive Enclosure
24	E7Y52A	HPE M6710 920GB SFF FE SSD
24	K0F27A	HPE M6710 1.92TB SFF FE SSD
1	TK821A	HP CS Rack Side Panel 1200mm Kit
1	TK816A	HP CS Rack Light Kit
1	TK815A	HP CS Rack Door Branding Kit
1	TK732A	HP 42U PDU MANAGEMENT-BRACKETS Cable
4	H8B50A	HPE Mtrd Swtchd 4.9kVA/L6-30P/NA/J PDU
1	BW932A	HPE 600mm Rack Stabilizer Kit
1	BW930A	HPE Air Flow Optimization Kit

Table 11-1 Continued.

Quantity	Part number	Description
1	BW891A	HPE Rack Grounding Kit
1	BD362A	HPE 3PAR StoreServ Mgmt/Core SW Media
1	BD365A	HPE 3PAR SP SW Latest Media
1	BD371A	HPE 3PAR App Suite for SQL Media
1	BD372A	HPE 3PAR App Suite for VMware Media
1	BD454A	HPE 3PAR OS Suite Current Media
27	D4T81A	HP StoreFront Analytics 3PAR LTU
1	H1K93A5	HPE 5Y Proactive Care 24x7 wDMR Service
1	H1K93A5#RZ6	HPE 3PAR 7440c/50OS Suite Base LTU Supp
1	H1K93A5#RZ7	HPE 3PAR7440c/50ReplSuite Base LTU Supp
1	H1K93A5#RZD	HPE 3PAR7440c/50Dynamic Optbaseltu Supp
1	H1K93A5#RZJ	HPE 3PAR7440c/50ReportingSuite LTU Supp
1	H1K93A5#RZP	HPE 3PAR7440c/50DataEncryption LTU Supp
64	H1K93A5#S7Y	HPE 3PAR7440c/50OS Suite Drive LTU Supp
64	H1K93A5#S7Z	HPE 3PAR7440c/50ReplSuiteDrive LTU Supp
64	H1K93A5#SDJ	HPE 3PAR7440c/50DynamicOptDriveLTU Supp
1	H1K93A5#SDR	HPE 3PAR 7440c/50 Priority Opt LTU Supp
64	H1K93A5#SDS	HPE 3PAR7440c/50PriorityOptDrv LTU Supp
32	H1K93A5#TR2	HPE 3PAR 7K 920GB SAS MLC FE SSD Support
27	H1K93A5#TR5	HPE StoreFront Anly VMw StoreServ Supp
1	H1K93A5#TRJ	HPE 3PARStoreServ7440c/50c 4N Base Supp
5	H1K93A5#WSF	HPE 3PAR Internal Entitlement Supp
6	H1K93A5#WUW	HPE 3PAR 7000 Drive Enclosure Support
4	H1K93A5#WUX	HPE 3PAR 7000 Adapter Support
32	H1K93A5#YL9	HPE M67101.92TB6GSAS 2.5 cMLC FESSD Supp
24	AJ837A	HPE 15m Multi-mode OM3 LC/LC FC Cable
5	C7536A	HP Ethernet 14ft CAT5e RJ45 M/M Cable
1	HA124A1	HP Technical Installation Startup SVC
1	HA124A1#5TL	HPE Startup 3PAR 7000 App Ste SQL SVC
27	K8G29AAE	HPE Ops Analytics for HPE OV E-LTU
27	M5R19A	HPE Cloud Optimizer for HPE OV LTU
27	700139-B21	HP 32GBmicroSDMainstream Flash Media Kit
1	H1K92A5	HPE 5Y Proactive Care 24x7 Service
27	H1K92A5#699	For HPE Internal Entitlement Purposes

Summary

This HPE Epic solution on CS700 was made operational at the factory and delivered in such a way that minimal customization needed to be done when the solution arrived on site. This solution provides the application user fast access and improved performance. Wrapping the various components with a rich and tightly integrated management suite provides a central point for day-to-day operations. Network, compute, storage, and hypervisor management were all fully integrated and tested together to give the administrator greater insight into the Epic solution and in turn provide a better user experience.

12 Hyper Converged System Running Multiple Workloads

INTRODUCTION

This chapter covers a municipality using HPE Hyper Converged System (Hyper Converged in this chapter) to host multiple workloads including Enterprise Resource Planning (ERP), Point of Sale (POS), and Microsoft SQL Server, on VMware ESXi Virtual Machines (VMs).

Hyper Converged was selected for this implementation because the municipality wanted a solution that is easy to setup, requires minimal administration, can provision VMs quickly, and is easy to maintain. These characteristics align perfectly with Hyper Converged pre-integrated systems that are factory-built to make setup and ongoing administration easy.

Solution overview

The ideal solution is a Hyper Converged HC 250 that has the following components (shown in Figure 12-1).

- 1 x HPE HC 250
- 1 x HPE OneView for VMware vCenter
- 2 x HPE Flex Fabric 5700

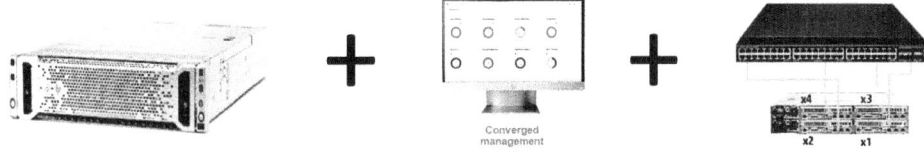

Figure 12-1 HPE HC 250 components

CHAPTER 12
Hyper Converged System Running Multiple Workloads

The HPE HC 250 combines compute, storage, and a pair of top-of-rack switches, plus OneView with RESTful API for all management and operational activities. Some key characteristics of this Hyper Converged design include:

- Easy to install, manage, provision VMs, and meet the other operational objectives listed earlier, including the following:
    o Instant on HPE OneView
    o 15 minute startup time
    o vCenter integration
- Software Defined Storage using HPE StoreVirtual VSA (called VSA in this chapter)
- High availability achieved by four redundant compute nodes and data protection in a single appliance
- Expansion up to 16 nodes
- Compact form factor that can accommodate up to 4 nodes in 2U
- Remote support using Insight Remote Support (IRS) to proactively scan for hardware failures
- Single phone number for support including one support ID for the entire solution, a dedicated support team to solve compute, storage, and application issues

Figure 12-2 shows the key components of the HPE HC 250 design example described in this chapter.

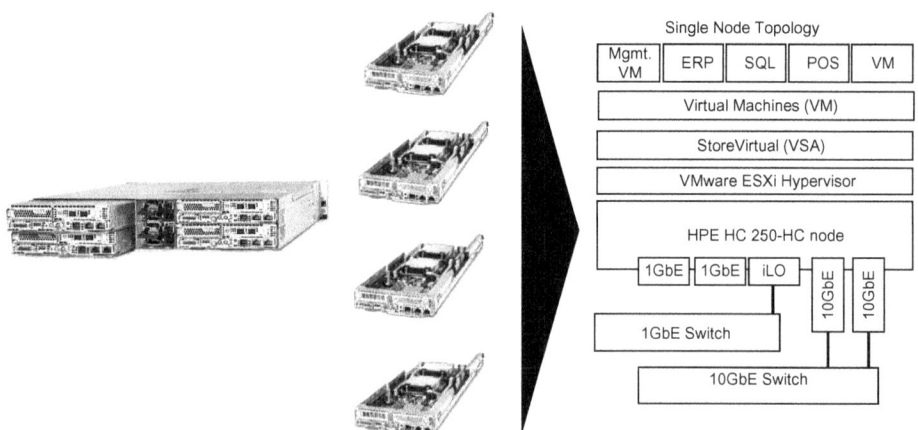

Figure 12-2 HPE HC 250 conceptual design

There are four nodes in the design in this chapter and all of the applications running in this diagram are virtualized. The next section covers the hardware inventory of the design.

Hardware inventory

Figure 12-3 shows the main components of the Hyper Converged solution that comprises up to 4 XL170r Gen 9 Servers.

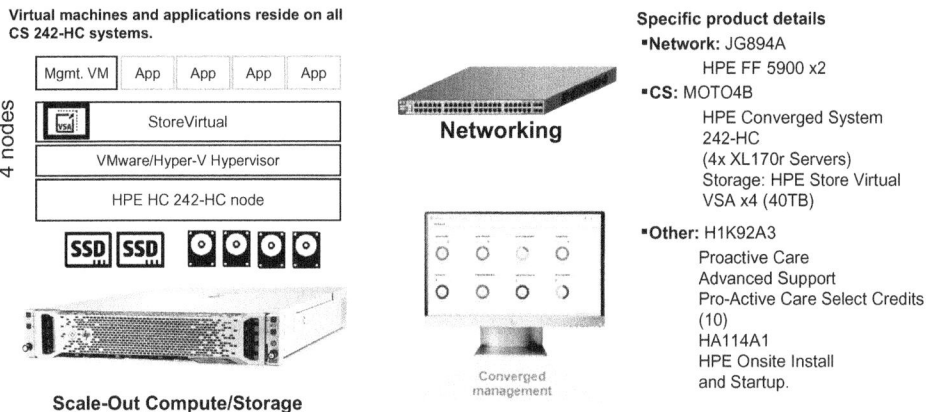

Figure 12-3 HPE HC 250 components

A significant benefit of the Hyper Converged system shown in Figure 12-3 is that it requires only a single part number for quoting and ordering. Hyper Converged systems are pre-installed and pre-configured with an all-inclusive feature set, including automated HPE optimization technologies that provide for easy installation and a faster time to production. To give you a more in-depth look at why a single SKU based appliance is preferable to a traditional Configure-To-Order (CTO), we will need to examine all of the components found inside the box. Table 12-1 breaks down the individual hardware components for both the HPE HC 250 chassis and the 4 individual nodes.

CHAPTER 12
Hyper Converged System Running Multiple Workloads

Table 12-1 Bill of materials for HC 250

QTY	Item #	Product description
1	M0T03B	HPE HC 250 System for VMware vSphere
1	795800-B21	HPE Z Apollo r2600 24SFF CTO Chassis
1	778792-B21	Zmod QR Label Apollo 2000
1	800097-B21	HPE ZLABELS Apollo 2000
1	802659-B21	HPE ZLBL AGNCY Apollo 2000—Oak PS
1	804266-B21	HZDOC Apollo 2000 WORLDWIDE
1	807635-B21	ZPKG 2U Apollo 2000 SFF UNIT
1	122657-00A	SOFTWARE TEST (US,M3 CTO)
1	106128-011	LBL, BLNK,THERMAL 4x6
1	M0T03-90032	Card, HPE Hyper CS250 VMware vSphere RTF
1	740713-B21	HPE t2500 Strap Shipping Bracket
1	822731-B21	HPE 2U Shelf-Mount Adjustable Rail Kit
1	852622-001	ASSY, BEZEL, StoreVirtual HC Gen9
2	800338-001	LBL, Win Server 2012 Embedded Std 2 CPU
1	AF728-84002	LBL, Warranty Entitlement
1	K2Q48-90802	MS EULA for WSS2012
2	800756-B21	HPE Z Apollo 2000 Fan Module
1	825048-002	LBL, HC 250-HC Hyper Converged
1	727259-B21	HPE Z 96W Megacell Battery 145mm Cable
1	798195-B21	HPE Z Apollo 2000 Server Node Blank Mod
1	654961-B21	HPE ZSFF HDD Common Blank Gen8
2	416151-B21	HPE ZPWR Cord 6FT C13 IEC To IEC
1	825037-B21	ZLBL, BIS Marks, RMN TPS-F021
4	M0T04B	HPE HC 250 Node
4	853319-B21	HPE ZXL1x0r Enhanced System Board
4	808659-B21	HPE ZXL1x0r Half-Width MB Tray
4	795806-B21	HPE ZBayonet Board XL170r
4	808660-B21	HPE ZBayonet Board XL170r
4	802672-B21	HPE ZBaffle XL170r
4	805609-B21	HPE ZLBL AGNCY XL170r Gen9
4	798179-B21	HPE ZXL170r Gen9 LP PClex16 L Riser Kit
4	798181-B21	HPE Z XL170r Gen9 LOMx8 R Riser Mod
4	798206-B21	HPE Z XL1x0r Gen9 Mini-SAS P840/440 Mod

(Continued)

Table 12-1 Bill of materials for HC 250—cont'd

QTY	Item #	Product description
4	798193-B21	HPE ZXL1x0r Gen9 DED IM Board Kit
4	726822-B21	HPE Z Smart Array P440/4G 12Gb Controller
4	784489-B21	HPE Z PCI to Controller 3Pin Cbl
4	784727-B21	HPE Z PCI to CNTRL 3 PIN LG Cbl
4	739483-B21	HPE Z Low Profile SA P430 Bracket
4	512514-B21	HPE Z iLO ADV NM 1 SRV 1YR TS&U LIC
4	808596-001	Phantom Agency Lbl XL170r Gen9
8	U20454-A00	CN:ACC:DUMMY PLUG RJ45 W/O KEY
4	805712-B21	ZHP AHCI-Enable FIO Setting
4	122657-00A	SOFTWARE TEST (US,M3 CTO)
4	864307-001	HPDS Diag Trigger
4	K2Q48-10577	HPE HC 250 for MS CPS Factory Image
4	825038-B21	ZLBL, BIS Marks, RMN TPS-F023
4	P9B51A	HPE HC 250 AW LTU for VMware vSphere 6.0
4	P9B51-90701	LTU,HPE HC 250 SW for VMware vSphere 6.0
4	5697-0862	LTU Entit - with Redemption
4	5697-3706	Universal LTU Envelope With Window
4	K2Q48-10595	HPE CS250-HC SV Management VM (6.0)
4	K2Q48-10603	HPE CS250-HC SV Factory Tools (6.0)
4	K2Q48-10591	HPE CS250-HC StoreVirtual VSA
4	K2Q48-10601	HPE CS250-HC SV VMware ESXi 6.0.0 Image
4	793024-L21	HPE XL1x0r Gen 9 Intel Xeon E5-2640v3 (2.6GHz/10-Core/25MB/105w) Processor
4	793024-L21	HPE XL1x0r Gen 9 Intel Xeon E5-2640v3 (2.6GHz/10-Core/25MB/105w) Processor
64	753222-B21	HPE 32GB 2Rx4 PC4-2133P-R Kit
4	665243-B21	HPE Ethernet 10Gb 2P 560FLR-SFP+ Adapter
16	781518-B21	HPE 1.2TB 12G SAS 10k 2.5in SC ENT HDD
8	816985-B21	HPE 480GB 6G SATA 2.5in MU-3 SFF SC SSD

CHAPTER 12
Hyper Converged System Running Multiple Workloads

As noted in Table 12-1, the HPE HC 250 consists of 4 individual compute nodes each with separate network connectivity. In order to simplify North/South and East/West traffic, a pair of HPE Flex Fabric 5700 switches were included for redundancy. Table 12-2 shows the various switch components.

Table 12-2 Switch components

QTY	Item #	Product description
2	JG894A	HPE FF 5700-48G-4xg-2QSFP+ Switch
4	JC680A	HPE A58xOAF 650W AC Pwr Supply
4	JC683A	HPE 58xOAF Frt (ports)-Bck (pwr) Fan Tray
8	JG330A	HPE X240 Qsfp+4x10GSFP+ 3m DAC Cable

Software inventory

To fulfill all of the business and operational requirements, this solution required several software applications to help coordinate the workload and ensure high availability across multiple compute and storage resources. Table 12-3 outlines the software components that were compiled for this solution.

Table 12-3 Software inventory

Item #	Product description	Additional details
1	ESXi 6.2	VMware Hypervisor virtualization system, creating virtual instances of workload applications. vCenter Management platform for daily provisioning and management.
2	MS OS 2012 R2 DC	Microsoft Operating System required to run each application
3	MS SQL 2014	Microsoft SQL 2014 required for data analytics and reporting, key component interacting with Federal Government ERP application.
4	HPE OV 4 vCenter	HPE OneView for VMware vCenter, responsible for providing seamless integration of hardware and software into a central management plane.
5	HPE VSA	HPE VSA Virtual San Appliance, tasked with data usage and retention, together with HPE OneView for VMware vCenter for single pane of glass provisioning and management.
6	HPE OV IO	HPE OneView Instant On, required for ease of startup and provisioning of hardware.
7	HPE IRS	HPE Insight Remote Support, tasked with automated remote support phone home calls

Implementation project plan

Due to the limited infrastructure requirements and technical skill set of the end user, this solution implementation needed to be well-organized and planned across various teams.

- HPE Technical Account Team: consulted with the user to develop a detailed Statement of Work, defining clearly the installation process.
- Factory Integration Services: pre-loaded the software applications, decreasing installation startup time.
- HPE Technical Services: racked the physical hardware and delivered installation/startup services, providing the onsite technical expertise for a worry free startup.

By researching the limitations of the physical infrastructure up front, the project team was able to build and deliver a solution that was tailored to the environment. The HPE Power Sizer Tool was used to validate that the solution would conform to the existing power and cooling infrastructure. Table 12-4 shows the solution's hardware power and cooling requirements.

Table 12-4 Power and cooling requirements for HPE HC 250

Rack level summary		Data center summary	
Line voltage	208 VAC	Line voltage	208 VAC
VA rating	536.55 VA	BTU HR	1824.89 BTU
BTU HR	1824.89 BTU	System current	2.58 A
System current	2.58 A	Total utilization input power	535.16 W
Utilization input power	535.16 W	VA rating	536.55 VA
Idle input power	157.08 W	Total idle input power	157.08 W
Max load input power	535.16 W	Total max load input power	535.16 W
System weight (Kg)	122.37 Kg		
System weight (lbs)	269.78 lbs		

These parameters were a key part of crafting the design and ensuring that it could be supported in several locations. The next section covers the validation and success criteria in more detail.

Solution validation and success criteria

A key factor in the success of this solution is determining the number of VMs that can safely run on the HC 250. Table 12-5 below shows the total amount of resources in the HC250.

CHAPTER 12
Hyper Converged System Running Multiple Workloads

Table 12-5 Total compute resources available

HPE HC 250 (Hybrid SAS+SSD)	Configuration options selection	Total resources available
# of Nodes in cluster	4 x XL170r Gen 9 nodes	4 x XL170r Gen 9 nodes
Physical CPU Cores	2 x Intel E5-2680v3/12 core @2.5GHz	96 physical cores
Available RAM in GB	4 x 512GB of RAM/node	2048 GB of RAM
Usable capacity in TB	4 x 1.2TB SAS 10k + 2 x 400GB SSD	6.8 TB of disk space

Unlike a physical deployment, you can create VMs whose total allocated resources can be more than are physically available in most host systems. In fact, oversubscription is a common practice in virtualized environments because most applications or workloads do not typically require their full allocated resources most of the time. These unused resources can be used by other virtual machines on the same host. The amount of oversubscription that is possible or acceptable will depend on your specific environment. Generally speaking, a virtual CPU (vCPU) to physical (pCPU) ratio of four or less (the total number of virtual CPUs from all of the VMs compared to the total physical cores in the hosts) and up to 125% of memory over-allocation should be acceptable for most typical workloads. In this example, CPU and memory allocations are less than what is available in the physical hosts. The total compute resources allocated and available for this solution is shown in Table 12-6.

Table 12-6 Total compute resources

4 SQL VM's required	SQL per VM sizing requirements	Total resource required
vCPU	4 vCPU's	16 vCPU's
RAM in GB	32 GB of RAM	128 GB of RAM
Disk capacity in GB	1000 GB of disk space	6000 GB of disk space
10 ERP VM's required	**ERP per VM sizing requirements**	**Total resource required**
vCPU	1 vCPU's	10 vCPU's
RAM in GB	30 GB RAM	300 GB RAM
Disk capacity in GB	30 GB of disk space	300 GB of disk space
6 POS VM's required	**POS per VM sizing requirements**	**Total resource required**
vCPU	2 vCPU's	12 vCPU's
RAM in GB	30 GB RAM	180 GB RAM
Disk capacity in GB	30 GB of disk space	180 GB of disk space
6 MGMT VM's required	**MGMT per VM sizing requirements**	**Total resource required**
vCPU	1 vCPU's	6 vCPU's
RAM in GB	10 GB of RAM	60 GB RAM
Disk capacity in GB	30 GB of disk space	180 GB of disk space

(Continued)

Table 12-6 Bill of materials for HC 250—cont'd

10 MISC VM's required	MISC per VM sizing requirements	Total resource required
vCPU	2 vCPU's	20 vCPU's
RAM in GB	30 GB RAM	300 GB of RAM
Disk capacity in GB	20 GB of disk space	200 GB of disk space
Total resources supplied	**Total resources used**	**Total resources available**
vCPU's	64 vCPU's	32 vCPU's
GB of RAM	968 GB of RAM	1080 GB RAM
GB of disk space	6860 GB of disk space	195 GB of disk space

As noted in this table, the total resources available can be customized for various applications thereby allowing for different resource requirements. For example, if your solution requires greater RAM per VM, then this could be solved by increasing the amount of RAM. As noted in Table 12-1, the configuration currently has a quantity of 64, 32GB DIMMS. For memory requirements the HPE 250 HC also supports 64GB and 128GB DDR4 Load Reduced DIMMs (LRDIMMs). Replacing the 32GB DIMMs with 64GB DIMMs will provide a total RAM of 4096 GB. Using all 28GB DIMMs will provide a total RAM of 8192 GB. There are also additional CPU models and disk configurations that are available and not shown in this chapter.

The next consideration is High Availability (HA). In a single appliance deployment, you will see the layout as shown in Figure 12-4.

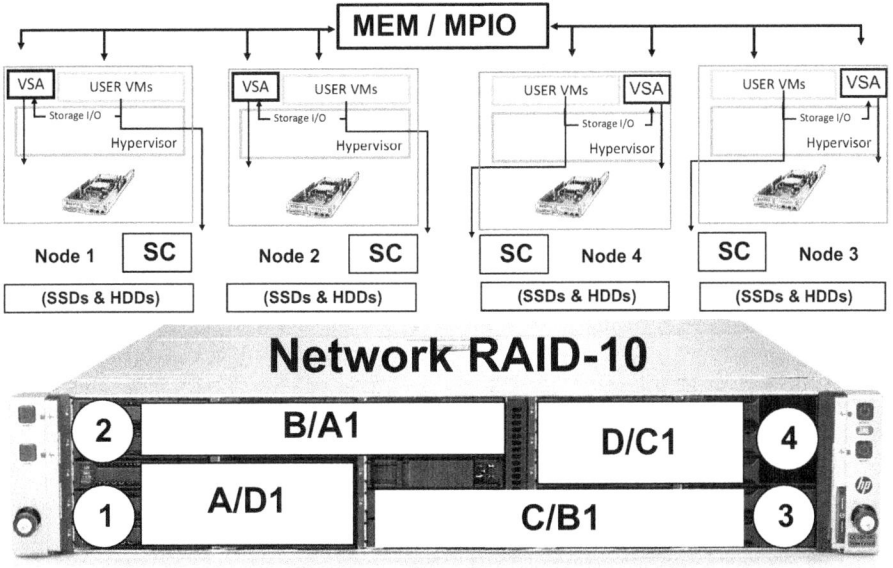

Figure 12-4 Network RAID-10

(VSA-Virtual Storage Appliance, VM-Virtual Machine, SC-Storage Controller, SSD-Solid State Drive, HDD-Hard Disk Drive, RAID-Redundant Array of Independent Disks)

Node-1 is the lower-left from the front of the appliance; Node-2 is in the upper left; Node-3 is in the lower right; and Node-4 is in the upper right. Using Network RAID-10 data can be read and written by any node to any node; however, each node controls its local data. Network RAID-10 is known as mirrored storage because two copies are written on two different nodes. Corrupt data on one node can be retrieved from another node using RAID-10.

When original or first data is written to Node-1 it is recorded as "A" data then replicated to Node-2. This would be denoted by VSA as "B/A1" meaning Node-1 is primary for this data and Node-2 secondary.

Shown at the top of Figure 12-6 are HPE StoreVirtual Multi-Path Extension Module (MEM) and Multi-Path IO (MPIO). MEM reduces the latency while improving and optimizing performance in ESXi environments through knowledge of the data location. This ensures Read I/Os are always serviced by the storage node that holds the authoritative data and Write I/Os are processed on the VSA Storage node which receives a duplicate copy. MPIO creates a fault tolerant solution through multiple I/O paths connecting each VM to storage. Network RAID-10 and MEM provides HPE VSA with 99.999% availability, creating data protection through multiple copies, and eliminates a single point of failure.

Summary

This solution provides a good example of the types of applications, including databases, which can be run on a Hyper Converged platform. The entire system can be preinstalled and preconfigured to enable the rapid provisioning of Virtual Machines. Many of the typical sizing and configuration considerations are covered in this chapter. Hyper Converged is now mainstream and immensely versatile. You will find solutions of this type running in central data centers, remote offices, branch offices, and locations where multiple workloads may need to be coordinated that integrate compute, storage, networking, and virtualization.

13 SAP HANA

INTRODUCTION

SAP HANA is an in-memory database that stores and retrieves data used by a variety of applications. This chapter will cover the design and implementation of a typical SAP HANA (called HANA in this chapter) nonproduction environment.

HANA can be deployed in a variety of ways including a purpose-built appliance or a more custom approach called Tailored Data Center Integration (TDI.) An appliance design for a nonproduction environment is covered in this chapter.

There are several components integrated in the HANA appliance including SAP HANA, SUSE Linux, and the advanced hardware components, in this case the hard partitioning capabilities of the system.

SUSE Linux Enterprise Server (SLES) is a supported and proven Operating System for SAP HANA and is used in this chapter. The following are some characteristics of SLES that make it ideal for this solution. SLES is:

- The leading Linux Operating Environment for SAP HANA
- Scalable and supports large quantities of processor cores and memory
- Supports SAP HANA replication process for High Availability Purposes
- Security focused – SUSE has developed the Operating System Security Hardening Guide for SAP HANA
- Used in numerous HANA benchmarks

SAP HANA deployment for nonproduction environment

This case covers a solution for a nonproduction environment using an HPE Converged System 900 (CS900) for SAP HANA. The CS900 uses an HPE Integrity Superdome X (called Superdome X in this chapter) as its compute block. The Superdome X is ideally suited for either scale-out or scale-up applications because it supports 24 TB of RAM, specifically for HANA, in a single instance and has a variety of features such as hard partitions to support a scale-out of multiple instances.

CHAPTER 13
SAP HANA

The project goal was to meet aggressive timelines to stand up a highly performant SAP HANA non-production development environment to support a modernization journey using the SAP Enterprise Resource Planning Software components.

The key objectives of this deployment were as follows:

- Quickly create an on-premises SAP HANA environment for nonproduction.
- Be well positioned for the near-term implementation of the production environment after this development phase.
- Partner with a solution provider who could provide experienced design, infrastructure, and implementation capabilities.

Sizing and design

Because HANA is part of an environment that is vital to the operation of an enterprise, sizing and design of the solution is important to the operation of this complex environment. Figure 13-1 depicts the sizing process at a high level.

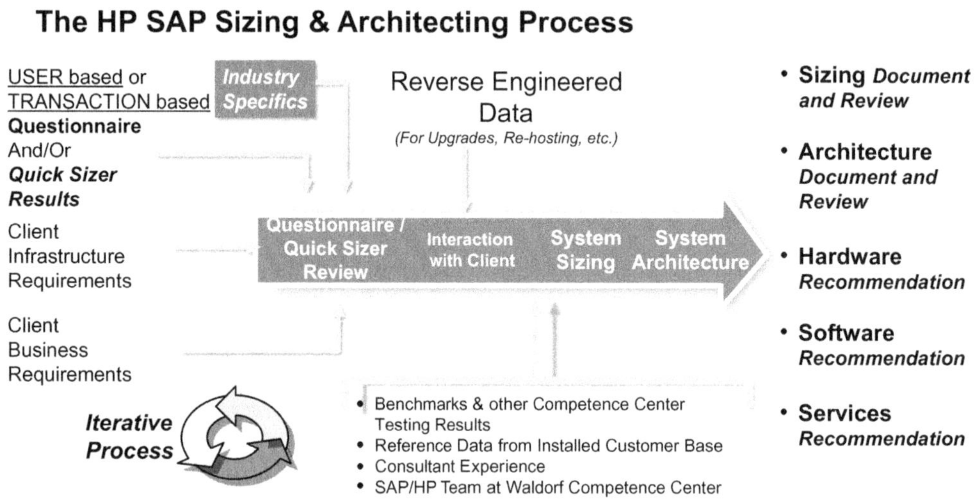

Figure 13-1 HPE SAP Sizing & Architecting Process

The flow of this diagram is to gather the key information about the requirements on the far left, work with various tools and processes in the middle, and then produce recommendations as shown on the far right.

>  **Note**
> This is a high-level diagram that does not depict the immense amount of complexity and effort required to produce the ideal HANA solution. HPE recommends consulting with HPE Technology Services for sizing and design of the solution.

Solution overview

This section describes the solution components and how we arrived at this solution.

SAP HANA infrastructure input sizing summary

Table 13-1 shows the parameters related to this HANA environment that is the basis for the sizing of this design.

Table 13-1 Sample SAP HANA sizing input

Environment	Sandbox (GB)	Development (GB)	QA (GB)
ECC	256	512	512
CRM	256	512	512
BW	256	512	512

The following is an explanation of the entries in this table, which provides the sizing requirement for the HANA instances:

Environments considered in scope for the modernization project include the following:

- ECC is SAP's ERP Central Component
- CRM is an integrated SAP customer relationship management solution
- BW is SAP's Business Warehouse

Landscapes for each of the environments:

- Sandbox is the area where ideas are experimented with or exercised for consideration in the development
- Development is the area for the Developers working on the specific environment
- QA is Quality Assurance, typically for unit testing
- Training is for the instance to support training

SAP infrastructure design assumptions

With the HANA sizing having been defined, the next step in crafting the design was to list the following assumptions:

- The development of nonproduction environment had to use the same appliance architecture as in production.

- In a typical HANA solution environment, there are also a variety of other non-HANA applications. Although non-HANA applications were not in scope, it was considered beneficial to the design that non-HANA applications typically could be added to the existing or refreshed x86 architecture.

Solution components

Based on the requirements gathered early in the process, a design was crafted for the HANA deployment that comprised the following components:

- CS900 that includes the following:
    - Superdome X Server
        - Provides the compute and memory engine for SAP HANA.
    - 3 PAR 8400 Storage
        - Used for persistent internal and dual purpose storage.
    - DL380 Management Server
        - Central Management Server, HP DL380 Gen9, contains components for SAP HANA studio, HPE 3PAR array, and other management software.
    - Management Network 5900 Top of Rack Switches
        - Consolidates all management connections and reduces the number of connections to the existing network.

Figure 13-2 depicts these components in the rack that was used for this environment.

Figure 13-2 shows that, although this is a complex solution, the hardware components consume a modest amount of rack space. A description of the components is shown to the right of the rack. The next section provides a more detailed breakdown.

TRIED AND TESTED SOLUTIONS TO ACCELERATE IT AND BUSINESS TRANSFORMATION
Ideal Platforms for Optimizing IT Workloads

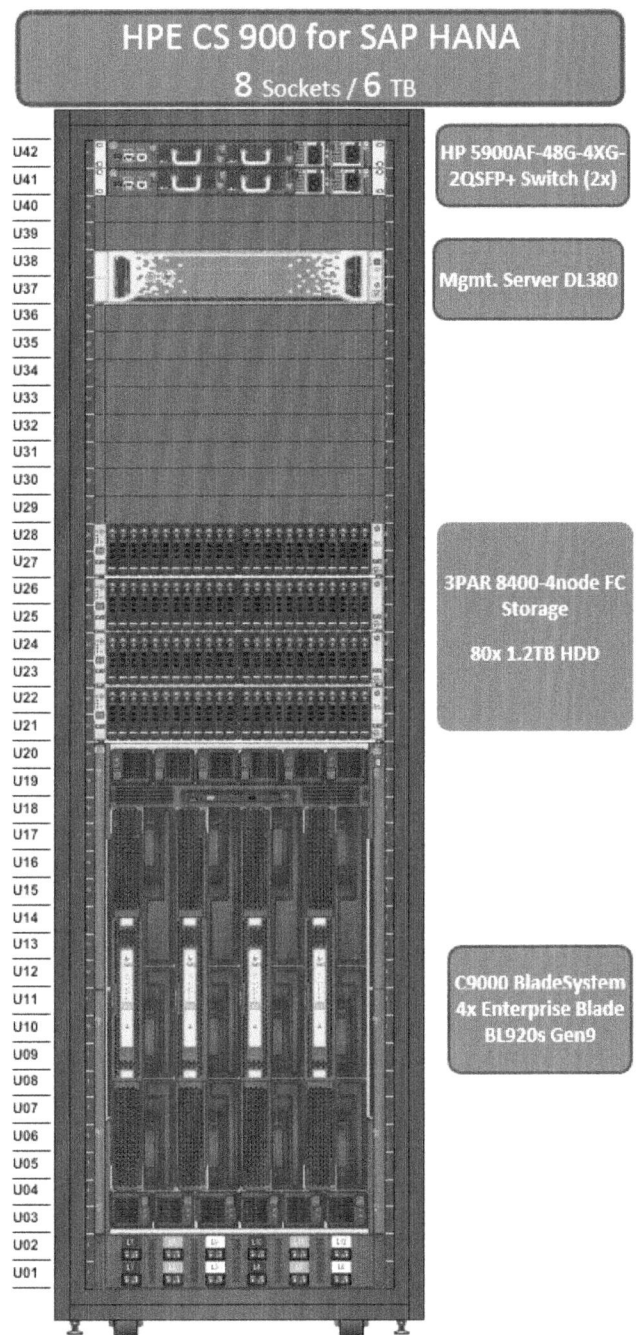

Figure 13-2 HPE CS900 configuration

CHAPTER 13
SAP HANA

Hardware inventory

Table 13-2 shows the Bill of Material (BOM) for the CS900 8-socket 6 TB configuration. Please note the current technology such as processor, memory, and so on, may have advanced since the writing of this document.

Table 13-2 Bill of Materials for CS900 Configuration

Quantity	Part number	Description
1	M0S67A	HPE 1075mm Shock Intelligent Rack
1	P9H67A	HPE SDX SAP HANA Scale-up Integratn Kit
1	P9H51A	HPE CS900 SAP HANA Scale-up Appliance
1	H8A03A1-420	HPE CS900 HANA Scale-up Rack FE Depl SVC
1	P9G87A	HPE 3PAR StoreServ 8400 4n Block
40	K2P93A	HPE 3PAR 8000 1.2TB SAS 10K SFF HDD
1	L7B69A	HPE 3PAR 8400 OS Suite Base LTU
80	L7B70A	HPE 3PAR 8400 OS Suite Drive LTU
1	BD362AAE	HPE 3PAR StoreServ Mgmt/Core SW E-Media
1	BD363AAE	HPE 3PAR OS Suite Latest E-Media
1	P9H73A	HPE SDX for SAP HANA Base Encl
2	787635-B21	HP 6127XLG Blade Switch Opt Kit
2	C8S47A	Brocade 16Gb/28c PP+ Embedded SAN Switch
4	P9H60A	HPE BL920s Gen9 E7-8880v4 44c Svr Blade
48	P9H75A	HPE SDX DDR4 128GB (4x32GB) Mem Module
8	P9H77A	HPE FF 20Gb 2p 650FLB Adptr for CS900
4	P9H78A	HPE QMH2672 16Gb FC HBA for CS900
4	J7J28A	HPE CS SAP HANA iLO Adv-BL 3yr TSU LTU
1	J7J23A	HPE CS SAP HANA Advanced Partitions LTU
1	H8B35A3	HPE 3Y Proactive Care Adv 24x7 Service
2	H8B35A3 9LS	HPE B-Series 8/24c Switch PowerPK Supp
2	H8B35A3 SPM	HPE 612x Blade Switch Support
4	H8B35A3 TPZ	HPE CS iLO Adv 3yr TSU LTU Support
1	H8B35A3 XWT	HPE SDX for SAP HANA Base Encl Supp
4	H8B35A3 XWU	HPE BL920s Gen9 Svr Bld Supp
4	N0U73A	SLES SAP 2Skt/1-2 VM 3yr 24x7 Flx LTU
2	P9H79A	HPE 5900AF 48G 4XG 2QSFP+ Swch for CS900
1	M0S75A	HPE SAP HANA DL380 Gen9 CMC Block
1	817927-L21	HPE DL380 Gen9 E5-2620v4 FIO Kit
1	817927-B21	HPE DL380 Gen9 E5-2620v4 Kit

Table 13-2 Continued.

Quantity	Part number		Description
8	805347-B21		HPE 8GB 1Rx8 PC4-2400T-R Kit
4	781516-B21		HP 600GB 12G SAS 10K 2.5in SC ENT HDD
2	781518-B21		HP 1.2TB 12G SAS 10K 2.5in SC ENT HDD
1	749974-B21		HP Smart Array P440ar/2G FIO Controller
1	727250-B21		HP 12Gb DL380 Gen9 SAS Expander Card
1	768896-B21		HP DL380 Gen9 Rear Serial Cable Kit
1	665243-B21		HPE Ethernet 10Gb 2P 560FLR-SFP+ Adptr
2	455883-B21		HPE BLc 10G SFP+ SR Transceiver
1	815868-B21		MS WS12 R2 Std FIO Npi en SW
1	H8B35A3		HPE 3Y Proactive Care Adv 24x7 Service
1	H8B35A3	S8X	HPE MS WS12 Standard OS+APP Support
1	H8B35A3	XW9	HPE SAP HANA DL380 Gen9 CMC Block Supp
1	BD505A		HPE iLO Adv incl 3yr TSU 1-Svr Lic
2	P9G88A		HPE 3PAR StoreServ 8000 SFF SAS Drv Encl
40	K2P93A		HPE 3PAR 8000 1.2TB SAS 10K SFF HDD
1	H8B35A3		HPE 3Y Proactive Care Adv 24x7 Service
1	H8B35A3	R2M	HPE iLO Advanced Non Blade - 3yr Support
2	H8B35A3	WSF	HPE 3PAR Internal Entitlement Supp
1	H8B35A3	XWA	HPE 3PAR StoreServ 8400 4n Block Supp
2	H8B35A3	XWB	HPE 3PAR StoreServ 8000 Drv Encl Supp
2	H8B35A3	XWV	HPE 5900AF Swch for CS900 Supp
4	H8B35A3	YMJ	HPE SLES SAP 1-2 VM 3yr Flx LTU Support
80	H8B35A3	YTV	HPE 3PAR 8000 1.2TB 10K SFF HDD Support
1	H8B35A3	YV1	HPE 3PAR 8400 OS Suite Base Support
80	H8B35A3	YV2	HPE 3PAR 8400 OS Suite Drive Support
6	C7536A		HP Ethernet 14ft CAT5e RJ45 M/M Cable
3	AF528A		HP 5xc13 PDU Extension Bars Kit
6	AF520A		HPE Intelligent 4.9kVA/L6-30P/NA/J PDU
50	U4VT2AS		HPE PCA Proactive Credits Per Year SVC
1	H1SR4AS		HPE TS Support Credits SVC
1	H0JD5A3		HPE 3Y TS Support Credits 30 Per Yr SVC
1	H0JD5A3	WFJ	HPE TS Support Credits 30 Per Yr SVC
1	H6K66A1		HPE Depl Accelerator for SAP HANA SVC
2	H8Q75A1		HPE Custom Consulting for SAP SVC
2	HL262A1		HPE Ancis 1-day Remote SVC

> **Note**
> In addition to these hardware components, many HANA-related components are part of the integration and deployment of the CS900 including the installation and configuration of the hard partitions of the Superdome X, the SUSE Linux Enterprise Servers (SLES), the networking layout, and finally SAP HANA itself. HPE recommends consulting with HPE Technology Services for onsite implementation.

Deployment overview

There were many steps involved in the planning, design, implementation, and support of this environment. Some detailed considerations are listed below:

Planning

- Obtain and install licenses for SAP HANA.
- Review, complete, and validate the SmartCID (Customer Intent Document) for each HANA instance. The SmartCID contains items such as networking identification and HANA DB components etc that enable the SAP HANA Appliance to be fully integrated and built at HPE.

Implementation

- Install the HPE CS900 for SAP HANA at the client site based on the SmartCID information.
- Connect the client's enterprise network for the CS900 iLO or OA.
- Validate and Install the required version of the SAP HANA instance for development.
- Verify and connect the client's enterprise network to the HANA instances on the CS900.

Knowledge Transfer

- Provide knowledge transfer on the use and operation of the CS900 and SAP HANA system
- Supply overview of SAP HANA hardware, configuration, and networking
- Supply overview of SAP HANA Studio and system administration

Summary

Although HANA is now mainstream, it is still a complex and mission-critical application that requires a tremendous amount of planning to ensure success. The example in this chapter features an appliance design for a nonproduction environment. Even though non-HANA applications are not part of the design, the choice of hardware architecture provides scope for expansion. This solution, like hundreds of others that have been designed and implemented by HPE Technology Services, had the right amount of planning to ensure a rapid and precise successful installation that met all of the customer's requirements.

14 Microsoft SQL Server Scale-Up Workload

INTRODUCTION

Large databases are ideal candidates to run on a scale-up platform. This chapter covers Microsoft SQL Server 2012 (SQL Server in this chapter) running on HPE Integrity Superdome X (called Superdome X in this chapter.) SQL Server 2016 is available at the time of this writing, but this deployment uses 2012. This is a large database environment that includes the following characteristics:

- 6 TB SQL Server Database size and growing exponentially.
- High CPU utilization (90%) of SQL Server DB on legacy Blade environment.
- New HPE 3PAR StoreServ 7400 array deployed for SQL Server replacing older EVA technology.
- I/O Throughput upgraded from legacy 4 GB FC to 16 GB FC.

Many customers are looking to modernize their Windows/SQL Server environments. SQL Server modernization provides immediate benefits to the environment including performance, scalability, and reliability.

The SQL Server workload described in this chapter realized these benefits and resulted in positive business outcomes.

Summary of Microsoft workload

This is a scale-up application that employs scalable Non-Uniform Memory Access (NUMA) architecture. In this case, like many scale-up workload designs, our goal was to meet the following criteria:

Business requirements

- Support rapid business growth with a more scalable solution beyond typical 2-4-socket x86 servers.
- Provide for future growth with the potential for the business to double in size.
- Assure timely order fulfillment to meet and exceed customer expectations.
- Increase workplace productivity by enabling employees to perform their jobs faster.

CHAPTER 14
Microsoft SQL Server Scale-Up Workload

Technical requirements

- Increase the number of batch requests per second.
- Speed query requests and end-user reporting.
- Optimize CPU utilization to improve SQL Server performance.
- Accelerate the inventory accounting process from 24–36 hours to 8 hours.

Solution overview

The basis for this scale-up solution is the HPE Superdome X platform. This server was selected because of its scale-up nature to meet the needs of high-end database demands. In addition, Superdome X has many mission critical features that are covered in Figure 14-1.

Scale on your terms:
Start small, grow seamlessly with your business

- Modular design with 1-8 Blades in Superdome X Enclosure
- Scale from 2-16 sockets of Intel E7v4 Broadwell Processors
- Provide CPU Core count from 8-384 in single Superdome X
- Scale memory from 512GB-24TB
- Support for Multi-OS including Microsoft Windows and SQL Server

Achieve breakthrough performance, even when you scale to the largest configurations

Figure 14-1 Superdome Scalability

The following are the details of the Superdome X configured in this solution:
- Qty (1) HPE Superdome X Server with the following characteristics:
- Qty (4) HPE Superdome X Server 2-socket Cellblades with (12) cores X 3.6Ghz processors.
- 1 TB RAM per Cellblade/ 4TB RAM total.
- Qty (4) Ethernet 10GB Dual-port 560FLB Network adapter.
- Qty (4) 16GB Fiber Channel SAN adapters.
- Qty (2) Brocade 16GB SAN switches.
- Qty (2) HPE 6125XLG Network switches.
- HPE Proactive Care Advanced Support for 3 year duration.

Figure 14-2 depicts the racks in the solution.

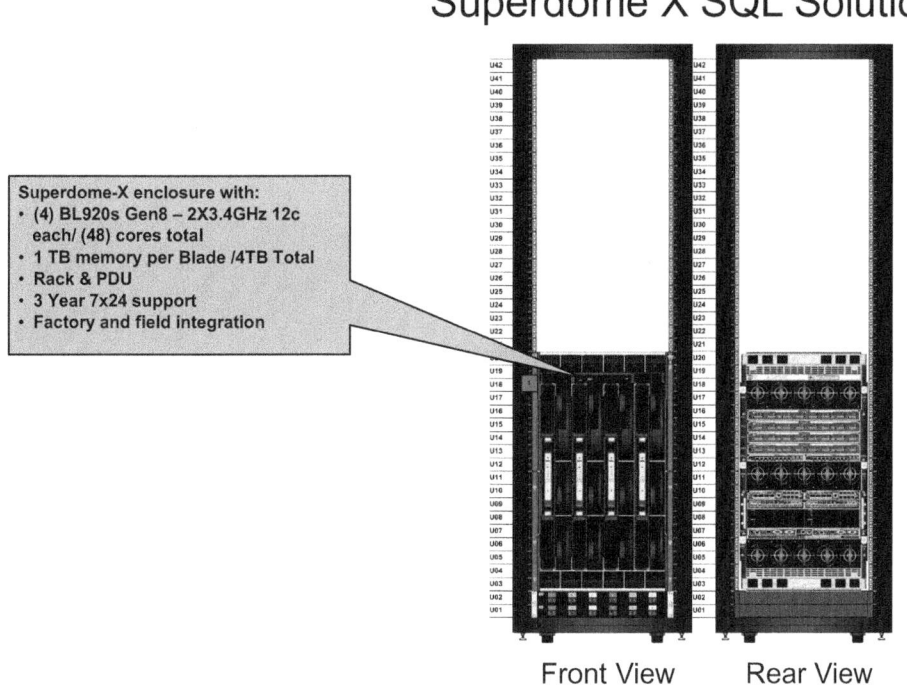

Figure 14-2 Rack Diagram of the Superdome X SQL Workload Solution

How did we arrive at this solution?

Substantial growth was expected by the firm deploying this solution, so ample room for expansion had to be built into the design. The functions supplied by this solution include the following:

- Warehouse Management
- Data Integration with key vendors
- Real-time order delivery verification to customers.

The previous SQL workload infrastructure was comprised of two HPE ProLiant BL685 full-height 4-socket server blades running SQL 2008 in a Microsoft active/passive cluster. This infrastructure was consistently running at 95% utilization and often bottlenecked at 100%. The result was a slow-down of business processing effecting order to delivery times significantly. The old architecture is shown in Figure 14-3.

SQL Server 2008 Legacy Architecture

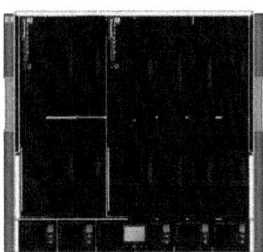

- Qty(2) BL685 G7 4-socket Blades in Microsoft Active/Passive cluster
- 4-socket 16 core AMD Opteron 6380 2.5 Ghzprocessors each
- 512GB DDR3-1333 Memory each
- 8GB FC SAN Infrastructure
- 1GbE Network Adapters
- Microsoft Windows and SQL Server

Figure 14-3 Legacy SQL Workload Architecture

The SQL modernization solution needed to address two keys areas. First, a modular solution was needed that allowed the SQL workload to scale with the business. Specifically, it needed a solution that could scale beyond 4-sockets and possibly to 8-sockets or beyond. Second, SQL Server is now licensed by the core count of a deployed server rather than the number of sockets. As a result, the SQL Server licensing and core count of the solution became very important sizing factors.

The Superdome X had the flexibility of offering several different Intel Xeon E7-v2 processors listed in Table 14-1.

Table 14-1 Superdome X E7-v2 Processor options

Name	Codename	Clock	Cores	Cache	TDP
Xeon E7-2890 v2	IvyBridge EX	2.8 GHz	15	37.5 MB	155 W
Xeon E7-2880 v2	IvyBridge EX	2.5 GHz	15	37.5 MB	130 W
Xeon E7-4830 v2	IvyBridge EX	2.2 GHz	10	20 MB	105 W
Xeon E7-8891 v2	IvyBridge EX	3.2 GHz	10	37.5 MB	155 W
Xeon E7-8893 v2	IvyBridge EX	3.4 Ghz	6	37.5 MB	155 W

 Note

Since the time that this solution was deployed, Superdome X processor technology has advanced to include Haswell (E7-v3) and Broadwell (E7-v4) offerings. These E7 processor technologies provide higher core counts and scalability and need to be evaluated based on SQL Server needs. Newer processor technologies are continually emerging.

Based on the existing BL685c solution with 2.5 GHz Processors, the Superdome X solution was architected based on the Intel Xeon E7-8893 v2 6-core IvyBridge processor.

In addition to the new 3.4 Ghz IvyBridge processors, there were several other infrastructure changes made with the Superdome X deployment including the following:

- Network interfaces on the Superdome X are 10 G NICs in comparison to the 1 GbE interfaces of the BL685c legacy solution.

- A new SQL Server storage solution was introduced into the environment. This was a HPE 3PAR 7400 with four nodes, 16 x 920GB Solid State Drives (SSD,) and 16 x 900GB 10 k RPM Hard Disk Drives (HDD)

- The Fiber Channel SAN infrastructure was upgraded to 16 GB FC including 4 X Qlogic 16 GB dual-port FC HBAs and a pair of 16 GB Brocade FC switches in the Superdome X Interconnect bays.

Hardware inventory

Table 14-2 shows the Bill of Materials (BOM) for this SQL Server workload solution. As discussed earlier, newer Intel E7 processors and other technologies for the Superdome X are available.

Table 14-2 BOM for SQL solution

QTY	Model #		Description
1	H6J66A		HPE 42U 600x1075mm Advanced Shock Rack
1	H6J66A	001	HP Factory Express Base Racking Service
1	HA455A1-000		HPE FE Solution Pkg 5 SVC
			SDX SQL Server
1	AT147A		Superdome X Enclosure
1	AT147A	002	
1	AT147A	0D1	
1	HA455A1-017		HPE FE Superdome X Pkg 5 SVC
1	AT152A		HPE Superdome X Advanced Par LTU
1	AT152A	0D1	Factory integrated
2	711307-B21		HP 6125XLG Blade Switch
2	711307-B21	0D1	Factory integrated
2	C8S45A		Brocade 16Gb/16c Embedded SAN Switch
2	C8S45A	0D1	Factory integrated
4	H7B42A		HP BL920s Gen8 3.4GHz 12c Svr Blade
4	H7B42A	0D1	Factory integrated
32	B9F09A		HP DDR3 128GB (4x32GB) Mem Module
32	B9F09A	0D1	Factory integrated
4	655639-B21		HP Ethernet 10Gb 2P 560FLB Adptr
4	655639-B21	0D1	Factory integrated

Table 14-2 Continued.

QTY	Model #		Description
4	710608-B21		HP QMH2672 16Gb FC HBA
4	710608-B21	0D1	Factory integrated
4	BD505A		HPE iLO Adv incl 3yr TSU 1-Svr Lic
4	BD505A	0D1	Factory integrated
1	H8B35A3		HPE 3Y Proactive Care Adv 24x7 Service
2	H8B35A3	85J	Brocade 4/12 and 4/24 SAN Switch Supp
4	H8B35A3	R2M	HPE iLO Advanced Non Blade - 3yr Support
2	H8B35A3	SPM	HPE 612x Blade Switch Support
1	H8B35A3	TQ0	HPE Integrity Superdome X Bas Enc Supp
4	H8B35A3	TQ1	HPE BL920s Gen8 Svr Blade Support
1	H6J85A		HPE Rack Hardware Kit
1	H6J85A	0D1	Factory integrated
1	BW906A		HPE 42U 1075mm Side Panel Kit
1	BW906A	0D1	Factory integrated
8	JD092B		HPE X130 10G SFP+ LC SR Transceiver
12	QK735A		HPE Premier Flex LC/LC OM4 2f 15m Cbl
4	AJ716B		HPE 8Gb Short Wave B-Series SFP+ 1 Pack
1	AT137A		HP SD2 Universal Rack Mounting Rail Kit
1	HA113A1		HPE Installation Service
1	HA113A1	5T3	HPE SD Server Rail Kit Installation SVC
1	HA124A1		HP Technical Installation Startup SVC
1	HA124A1	5VH	HPE Startup Integrity Superdome X SVC
10	U4VT2AS		HPE PCA Proactive Credits Per Year SVC

Software inventory

There are several software components that are essential for the operation of this SQL Server workload. Table 14-3 shows the software components that are implemented in this SQL solution.

Table 14-3 Software used to implement this SQL environment

Item #	Product description	Additional details
1	Windows 2012R2	Windows operating system with custom implementation.
2	SQL Server 2012	Microsoft SQL Server database for business processing.
3	Microsoft QFEs	Microsoft Quick Fix Engineering (QFEs) provides service updates (hotfixes) to Microsoft products.
4	Superdome X IO Firmware and Drivers Image	Drivers and Firmware specific to Superdome X server platform. Provided by HPE.
5	Windows WBEM Management	Supply system management data for the health and monitoring of the Superdome X Windows server. Provided by HPE.

Implementation plan

Table 14-4 lists the high-level tasks performed to implement this solution.

Table 14-4 Simplified Implementation Plan for SQL Workload Solution

Datacenter Power and Rack Prep complete
Superdome X Arrival at third party Datacenter and moved to appropriate location
Superdome Hardware Installation (rack and power up)
Configure and Verify Superdome X iLO Management Port connectivity
Superdome X firmware updated and verified
Superdome X Hard Partition configuration (Two Npars) and verification.
Superdome X 16GB SAN switch configuration
Superdome X 6125 Network Switch configuration
Customer provides network cabling and completes any changes to the Juniper network core.
Windows 2012 R2 installations (iso image provided by customer)
Superdome-X specific drivers, providers, hotfixes and best practices installed on both nPars
Microsoft Cluster Services installed between Npartitions
SQL Server 2012 Enterprise Edition installation on both Npartitions
HPE Customer Engineer to run Superdome X Hardware Diagnostic scripts and analyze
HPE to verify Windows Network Teaming and FC MPIO configuration is correct and functioning
SQL Server 2012 Testing
Final Production Deployments

This simplified implementation plan had several subtasks for each of the high-level tasks shown. The first activity, for instance, includes data center power which in some cases may require additional rack power installed and may need to be done well in advance of the equipment arriving on-site. In this installation, a project manager was assigned to ensure that the entire process progressed smoothly.

The next section covers some of the expertise and skills required in this installation.

Expertise and skills

There were experts involved in architecting this SQL Server solution including, but not limited to, a team that consisted of people with the following skills:

- SQL Server database expertise—The performance and scaling of the SQL Server database was a critical technical success factor to the SQL Server Modernization. The experts engaged in many project phases had SQL Server expertise and a strong understanding of the current platform performance.

- Architecture development—The knowledge of the legacy SQL Server environment was a key to sizing the new Superdome X-based Solution. There was no load testing performed, so the empirical data from the legacy environment was crucial to proper sizing. A team with knowledge of legacy SQL Server environment and Superdome X performance crafted this design.

- Implementation team— Once functional testing was complete, the production environment was implemented quickly by a team with infrastructure, SQL Server, and application hands-on experience.

- Project management—The key to this implementation, due to short time frames for migration, was a detailed and coordinated project plan. A successful migration from legacy environment was only possible with project management.

The next section covers the validation of the installation including a performance problem that was uncovered and rectified in order to achieve optimal performance.

Solution validation and success criteria

The final Superdome X configuration was to put in place in a third-party data center for initial configuration in advance of testing and production deployment.

Due to the critical nature of this business, the cutover was carefully planned to minimize downtime to the server and SQL Server database. The full cutover consumed roughly 70 hours over a long weekend.

Application architecture

The SQL Server workload needed specific characteristics in order to properly perform in a scale-up environment including the following:

1. CPU scalability including large multi-core processors and 4-socket and beyond servers
2. Large memory footprint in the range of 2-4 TB and beyond
3. Network throughput of 10 Gb minimum for application processing, data integration, and backup.
4. A highly reliable server platform with minimal downtime.

The Superdome X architecture was capable of providing these characteristics of SQL workload including a unique set of Reliability, Availability, and Serviceability (RAS) features.

Application testing

After the initial deployment, the SQL Server database testing consisted of simple SQL calls to verify the results on the SQL Server workload on the new Superdome X platform. These initial tests, although not specific to load testing, were successful. As time progressed, however, response time began to slow and analysis of the performance issue took place.

Performance issue

As mentioned earlier, the Superdome X is a NUMA server capable of very scalable SQL workloads. The Superdome X 2-socket modular blades referred to as cellblades can be combined into a single SMP architecture to scale a SQL Server workload. Each individual cellblade represents a NUMA set or node of the environment. This particular architecture consists of 6-core processors resulting in 12-core Superdome X cellblades as depicted in Figure 14-4.

CHAPTER 14
Microsoft SQL Server Scale-Up Workload

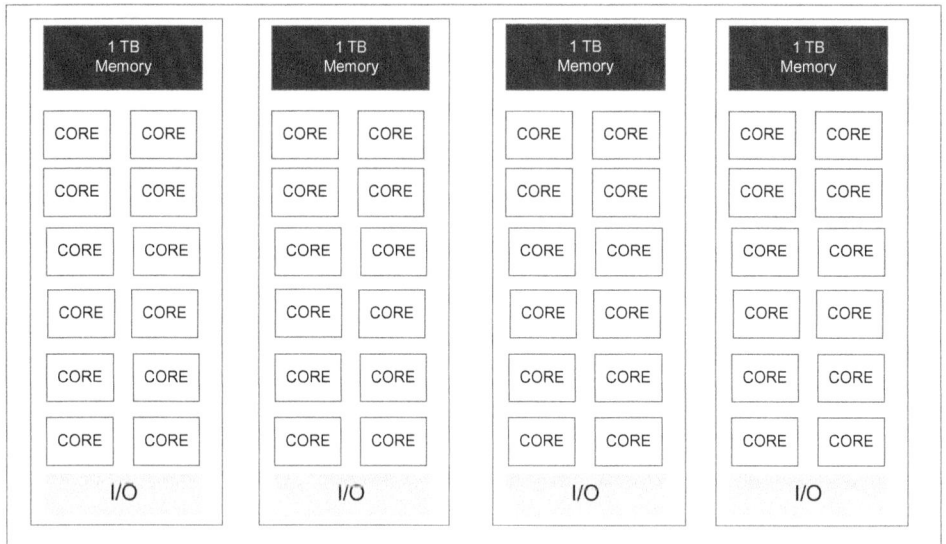

Figure 14-4 Superdome X NUMA architecture

The result is scalable 8-socket, 48 core, and 4 TB of memory SQL Server database instance environment which consists of four NUMA sets or nodes.

The SQL Server version deployed for production was SQL Server Enterprise Edition. This was the database edition that had run on the previous BL685 blade architecture and had not changed in an effort to minimize change to the environment.

This version of SQL Server has a licensing limitation of 40 processors. As a result, SQL Server would only recognize 40 cores in the environment. The result was a default distribution of cores across the NUMA sets as depicted in Figure 14-5.

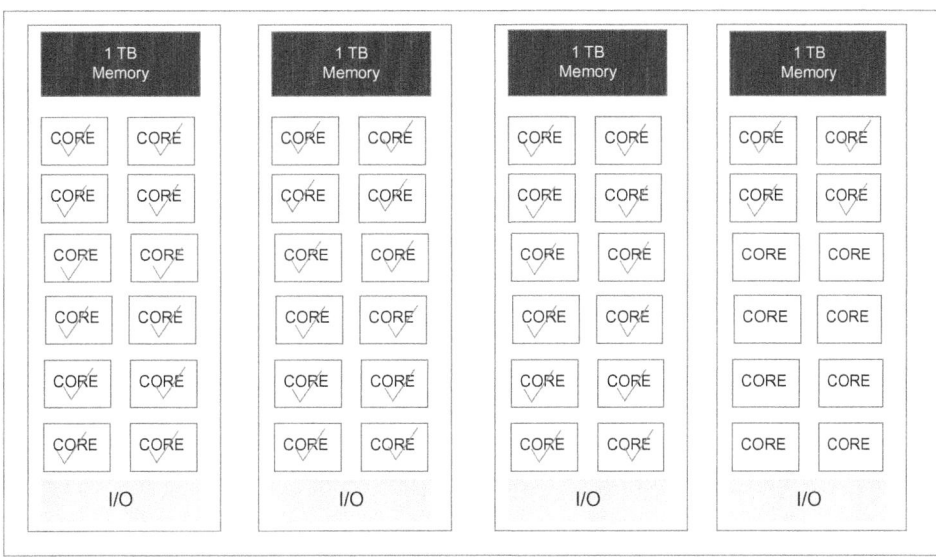

Figure 14-5 Superdome X with unequal CPU distribution within NUMA sets

SQL Server will distribute the workload evenly across the NUMA sets. The result in this particular case was that NUMA sets 1–4 were equally loaded. However, NUMA set #4 has one-third the processing power (4 cores) in comparison to the other NUMQ sets (12 cores each).

The result of this NUMA configurations was that some SQL workloads when scheduled for NUMA set #4 were significantly slower as a result. Once this was discovered, the NUMA configurations were balanced as depicted in Figure 14-6 and the performance improved dramatically.

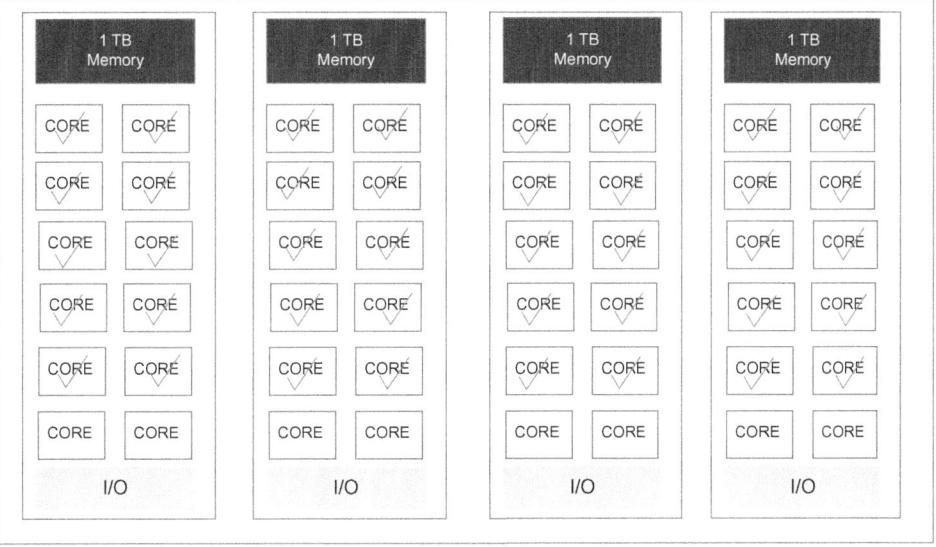

Figure 14-6 Evenly distributed NUMA sets

Analysis

After the analysis and resolution of the early production SQL Server performance issues, the workload was performed at significantly less CPU utilization. The overall CPU utilization was now 34% consistent.

The result of the low CPU utilization is the guaranteed performance during the peak periods of the year. In addition, the SQL Server workload on Superdome X has proven to be the growth platform as the business grows significantly.

The current Superdome X server can grow 2X in size with the current CPUs. Any growth of the SQL Server workload beyond 2X can be accomplished with an additional server. This would essentially split the Microsoft cluster across physical servers. In addition, growth could also be accomplished by increasing the performance density with Intel E7 v3 (18 cores each) or E7-v4 processors (24 cores each) that are available for the Superdome X Cellblades today.

Summary

The deployment of the SQL Server workload on Superdome X was a relatively simple process. The Superdome X was delivered with the components (Cellblades, LAN/SAN Switches, and Onboard Administrators modules) fully integrated into the Superdome X chassis via HPE Factory Express service. In addition, these components are run through a series of diagnostics at the HPE factory to ensure the server is free of any issues or defects. Superdome X employs electrically isolated partitions called nPartitions. The Windows Server OS boot drives for both nPartitions were placed on the 3PAR 7400 on RAID protected drives, removing the need for a separate local boot array.

This solution was deployed very quickly. However, the performance issued described earlier needed to be resolved with a team of experts in various disciplines. This is a complex scale-up application and intimate knowledge of the environment was required to resolve the problem. As with many complex solutions, problems will inevitably be encountered and so the right set of experts are needed to ensure mission critical applications get up and running quickly.

15 Backup and Recovery of Virtual Machines

INTRODUCTION

There are a variety of ways to deploy backup and recovery of virtual environments and there exist countless tools and products that can be used for backup and recovery. This chapter covers one successful technique that employs HPE StoreOnce 4500, Veeam® software, and some other components. The following is a high-level summary of this solution:

- HPE StoreOnce 4500 Backup
- HPE StoreOnce Catalyst
- Veeam® Backup software-version 9.0
- High-performing backup with a short Recovery Point Objective (RPO)
- Support for both fiber channel and iSCSI.

Solution overview

In order for this StoreOnce-Veeam® solution to provide predictable recovery times and reliable data protection, both fiber channel and iSCSI connectivity are employed. Fiber channel enables the creation of a fiber channel fabric, which acts as the core component of the Storage Area Network (SAN). Internet Small Computer System Interface (iSCSI) is used to link data storage devices over a network to transfer data. It does so by transmitting SCSI commands over an Internet Protocol (IP) network. iSCSI uses a Gigabit Ethernet (Gbe) interface at the physical layer, which allows this system to connect directly to the Gbe switches and/or IP routers that are used in this solution.

CHAPTER 15
Backup and Recovery of Virtual Machines

Figure 15-1 shows the key components and processes of the StoreOnce-Veeam® solution.

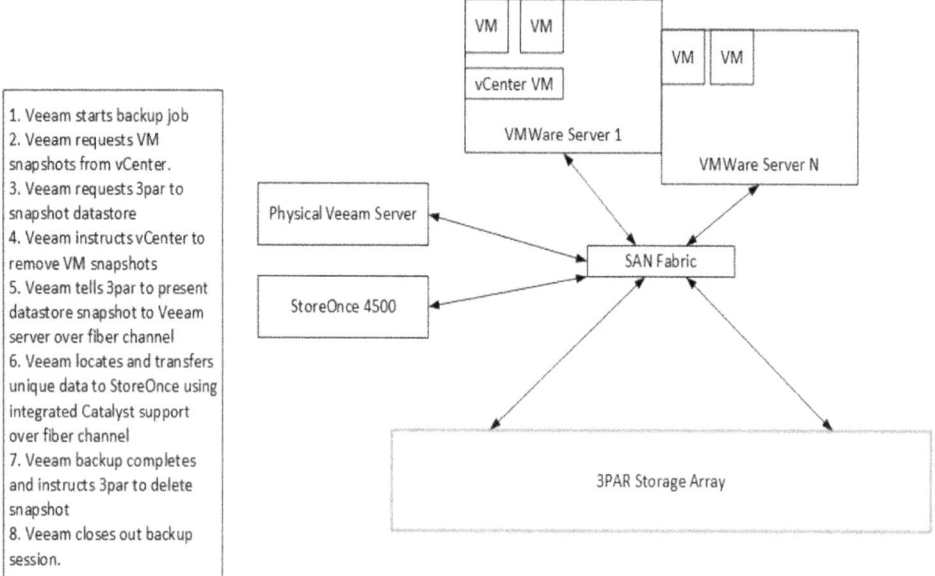

Figure 15-1 Components of the StoreOnce-Veeam backup solution

Figure 15-2 shows the rack diagram.

Ideal Platforms for Optimizing IT Workloads

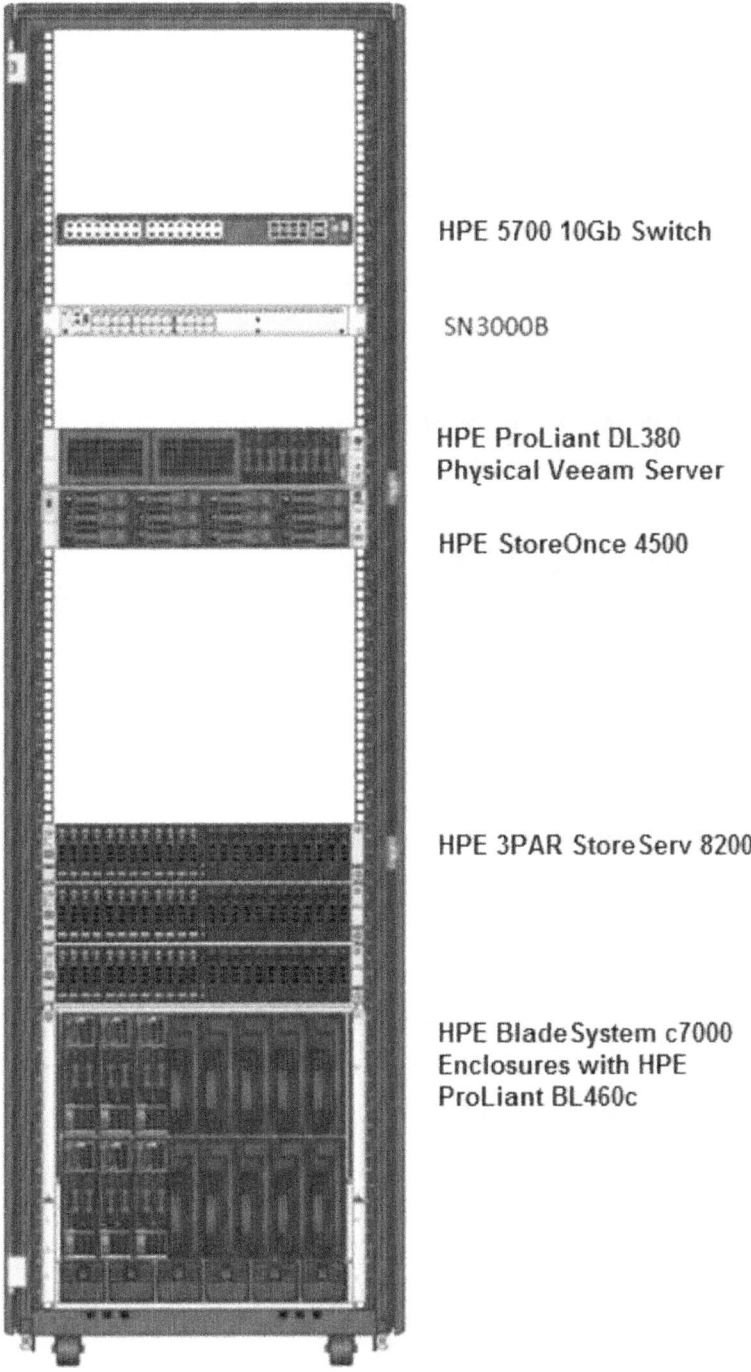

Figure 15-2 Rack Diagram depicting the scale-out solution

CHAPTER 15
Backup and Recovery of Virtual Machines

The following components are included in this solution:

- HPE StoreOnce 4500 Disk Backup System
- HPE StoreServe 8200 SAN Storage Array
- HPE ProLiant DL380 Server
- C7000 Chassis with Virtual Connect 20/40
- BL460 g8 blade servers
- HPE Fiber channel switch
- HPE 10 GBe Ethernet switch
- HPE StoreOnce Catalyst Licensing
- Veeam® V9.0 Enterprise Plus Edition licensing

The environment consists of six Proliant blades running VMware® v6.0 to support production Microsoft SQL Server®, Citrix®, proprietary applications, and general user support. This solution is for roughly 16–18 TB of useable data space.

How did we arrive at this solution?

The requirements for this solution had the following characteristics:

- Low tolerance for lost data
- Continually expanding data capacity that required backup
- Minimal downtime due to the critical nature of the applications that are included in the backup solution

The StoreOnce-Veeam® solution addresses the challenges of Recovery Point Objective (RPO) and Recovery Time Objective (RTO).

Figure 15-3 shows performance numbers attained using Veeam® Backup and Recovery with the StoreOnce 4500 in this environment. This example shows that we achieved a 440 GB backup in 34 minutes at 861 MB/s. The StoreOnce could actually perform faster because there is a bottleneck noted at the source (Storage). Adding additional drives to the storage would potentially add increased backup performance.

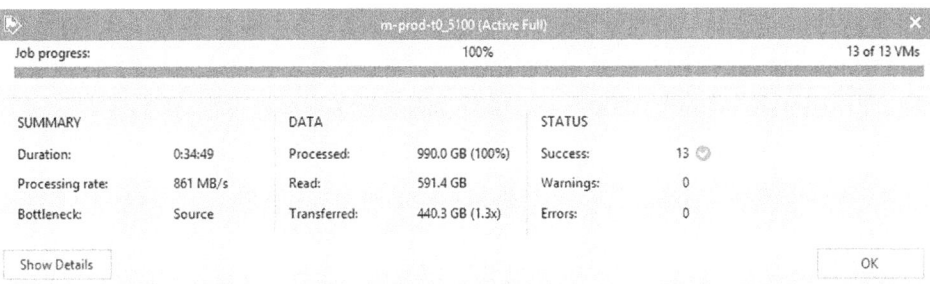

Figure 15-3 Sample Backup Performance Numbers

Hardware inventory

Table 15-1 shows the hardware Bill of Materials (BOM) for this backup and recovery solution. There was a sizing exercise that took place to ensure that the capacity was sufficient for current and short-term growth need of the environment.

Table 15-1 BOM for scale-out solution

Quantity	Model #		Description
0			**HP DL360 Gen9 E5-2630v3 Base SAS Svr [#1]**
1	755262-B21		HP DL360 Gen9 E5-2630v3 Base SAS Svr
1	726719-B21		HP 16GB 2Rx4 PC4-2133P-R Kit
4	781518-B21		HP 1.2TB 12G SAS 10K 2.5in SC ENT HDD
2	AK344A		HPE 81Q PCI-e FC HBA
1	H6J66A		HPE 42U 600x1075mm Advanced Shock Rack
1	H6J66A	001	HP Factory Express Base Racking Service
0			**Complex 1/Blade C7000 BL460 Blades**
1	681844-B21		HP BLc7000 CTO 3 IN LCD Plat Enclosure
1	E5Y41A		HPE OV 3yr 24x7 Encl FIO Phys 16 Svr Lic
6	727021-B21		HP BL460c Gen9 10Gb/20Gb FLB CTO Blade
6	726997-L21		HP BL460c Gen9 E5-2609v3 FIO Kit
24	726718-B21		HP 8GB 1Rx4 PC4-2133P-R Kit
12	781518-B21		HP 1.2TB 12G SAS 10K 2.5in SC ENT HDD
6	700764-B21		HP FlexFabric 20Gb 2P 650FLB FIO Adptr
6	761871-B21		HP Smart Array P244br/1G FIO Controller
6	651281-B21		HP QMH2572 8Gb FC HBA
2	AJ820B		HP B-series 8/12c BladeSystem SAN Switch
2	638526-B21		HP BLc VC Flex-10/10D Module Opt

CHAPTER 15
Backup and Recovery of Virtual Machines

Table 15-1 Continued.

Quantity	Model #	Description
4	AJ716B	HPE 8Gb Short Wave B-Series SFP+ 1 Pack
6	733459-B21	HPE 2650W Plat Ht Plg Pwr Supply Kit
6	412140-B21	HP BLc Encl Single Fan Option
1	456204-B21	HP BLc7000 DDR2 Encl Mgmt Option
1	433718-B21	HP BLc7000 10K Rack Ship Brkt Opt Kit
1	677595-B21	HP BLc 1PH Intelligent Power Mod FIO Opt
		StoreOnce 4500 HPE Disk Backup System
1	BB878A	HP StoreOnce 4500 24TB Backup
1	H6J85A	HPE Rack Hardware Kit
2	H5M60A	HPE Basic 8.3kVA/CS8265C/C13 C19/NA PDU
1	BW932A	HPE 600mm Rack Stabilizer Kit
1	BW932A B01	Include with complete system
1	BW930A	HPE Air Flow Optimization Kit
1	BW930A B01	Include with complete system
1	BW906A	HPE 42U 1075mm Side Panel Kit
1	755384-B21	HP DL360 Gen9 E5-2630v3 Kit
1	BB888A	HPE StoreOnce 4500/5100 Catalyst LTU
1	BD505A	HPE iLO Adv incl 3yr TSU 1-Svr Lic
1	P8B24A	HPE OV w/o iLO 3yr 24x7 Phys 1Svr LTU
7	Q0J93AAE	HPE Docker CS Eng Serv Bndl 7x24 1Yr Sub
8	QK734A	HPE Premier Flex LC/LC OM4 2f 5m Cbl
		Fiber Channel Switches
2	AM868C	HPE 8/24 Base 16-port Enabled Switch
2	AM868C 05Y	2.4m Jumper (IEC320 C13/C14, M/F CEE 22)
2	TC472AAE	HPE Intelligent Inft Anlyzer SW v2 E-LTU
32	QK735A	HPE Premier Flex LC/LC OM4 2f 15m Cbl
32	AJ716B	HPE 8Gb Short Wave B-Series SFP+ 1 Pack
		3Par StoreServe 8200
1	K2Q36A	HPE 3PAR StoreServ 8200 2N Fld Int Base
2	H6Z00A	HPE 3PAR 8000 4-pt 16Gb FC Adapter
8	K2P94A	HPE 3PAR 8000 1.8TB SAS 10K SFF HDD
4	K2R27A	HPE 3PAR 8000 1.92TB SAS cMLC SFF FE SSD
1	L7B45A	HPE 3PAR 8200 OS Suite Base LTU

Table 15-1 Continued.

Quantity	Model #	Description
1	L7B45A 0D1	Factory integrated
32	L7B46A	HPE 3PAR 8200 OS Suite Drive LTU
1	L7B47A	HPE 3PAR 8200 Data Opt St v2 Base LTU
32	L7B48A	HPE 3PAR 8200 Data Opt St v2 Drive LTU
1	L7D49A	HPE Smart SAN for HPE 3PAR 8xxx LTU
2	E7Y71A	HPE 3PAR 8000 SFF(2.5in) Fld Int Dr Encl
16	K2P94A	HPE 3PAR 8000 1.8TB SAS 10K SFF HDD
4	K2R27A	HPE 3PAR 8000 1.92TB SAS cMLC SFF FE SSD
1	BD362AAE	HPE 3PAR StoreServ Mgmt/Core SW E-Media
1	BD363AAE	HPE 3PAR OS Suite Latest E-Media
12	QK734A	HPE Premier Flex LC/LC OM4 2f 5m Cbl

Software inventory

Software components are equally important for a complete backup solution, and these are shown in Table 15-2.

Table 15-2 Software used to implement data protection environment

Item #	Product description	Additional details
1	StoreOnce Catalyst	Required for each StoreOnce device
2	Veeam® Backup Software	V9.0 Enterprise Edition is required
3	VMware®	V6.0 Enterprise Plus edition

Implementation steps

The step-by-step instructions included in the following sections are broken down into the following two high-level processes:

- Fiber channel instructions
- iSCSI instructions

The environment includes the implementation of StoreOnce-Veeam® that includes an HPE StoreServ 8200 connected via a fiber channel switch connected to an HPE ProLiant DL380 server. The server runs Veeam® V9.0 in a nonvirtualized server for performance considerations. StoreOnce 4500 is connected to both the fiber channel and iSCSI environments.

There is also an HPE BladeSystem C7000 Blade Enclosure chassis with the HPE ProLiant BL460's servers running VMware® 6.0 with a variety of applications running in the virtual machines.

Fiber channel step-by-step setup instructions

The following are step-by-step instructions for the fiber channel setup:

1. Log onto to the StoreOnce 4500 device.
2. Navigate to the StoreOnce 4500 user interface (UI) and expand it.
3. Select Option in StoreOnce Catalyst.
4. Select on the Fiber channel settings tab.
5. Record the unique identifier that displays at the upper middle segment of the screen. The unique identification number starts with "COFC."
6. Record the World Wide Name (WWN) for the StoreOnce device. Zone to the host bus adapter in the backup server.
7. Navigate to the StoreOnce 4500 pane and select Catalyst Stores
8. Create a new Store by selecting on Create
9. Name the Store
10. Change the primary and secondary policy to low bandwidth
11. Select Create

The fiber channel is now properly setup. Log out of the StoreOnce 4500 Interface.

Now open the Veeam® V9.0 interface.

1. Select the backup infrastructure tab on navigation bar lower left of the window
2. Select Add Repository
3. Name the repository within Veeam®
4. Select Next
5. Select Deduplication Appliance
6. Select HPE StoreOnce
7. Select Next
8. Select "Use fiber channel connectivity"
9. Type the unique identifier of the StoreOnce device gathered during the prior steps
10. Provide the StoreOnce credentials
11. Select the Gateway server that is zoned to the StoreOnce device

12. Select Next
13. Enter the Maximum Concurrent Tasks = Qty XX
14. Select Next
15. Keep "Mount Server" as default
16. Select Next

    Veeam® provides a summary page for the configuration created. At this point, Veeam® may install vPower Network File System (NFS)
17. Select Next
18. Select Finish

iSCSI step-by-step setup instructions

1. Log into the StoreOnce 4500 device
2. Navigate to StoreOnce and expand the UI
3. Select Option in StoreOnce Catalyst
4. Navigate to the StoreOnce pane and click on Catalyst Stores
5. Select a new Store by selecting the Create button
6. Name the Store
7. Change the primary and secondary policy to low bandwidth
8. Select on Create

The ISCSI is now properly setup. Log out of the StoreOnce 4500 interface.

Now open the Veeam® V9.0 interface.

1. Select on backup infrastructure tab on navigation bar on the lower left of the Veeam® window
2. Select Add Repository
3. Name the repository within Veeam®
4. Select Next
5. Select Deduplication Appliance
6. Select HPE StoreOnce
7. Select Next

CHAPTER 15
Backup and Recovery of Virtual Machines

8. Type in the IP address or FQDN of the StoreOnce 4500 device
9. Provide the StoreOnce credentials
10. Select the Gateway server that is connected to the StoreOnce 4500 device
11. Select Next
12. Enter the Maximum Concurrent Tasks = Qty XX
13. Select Next
14. "Mount Server" should be left as default
15. Select Next

    Veeam® provides a summary page for the configuration created and may install Veeam® vPower Network File System (NFS).

16. Select Next.
17. Select Finish.

At this point, verify that you can perform either a Synthetic Full or an Active Full for every six incremental backups when configuring your Backup job.

Veeam® is a prerequisite to implementing this process. Figure 15-4 provides an example of what you will see from the Veeam interface to demonstrate the kind of screen you should be viewing. Veeam® is used to manage the backup.

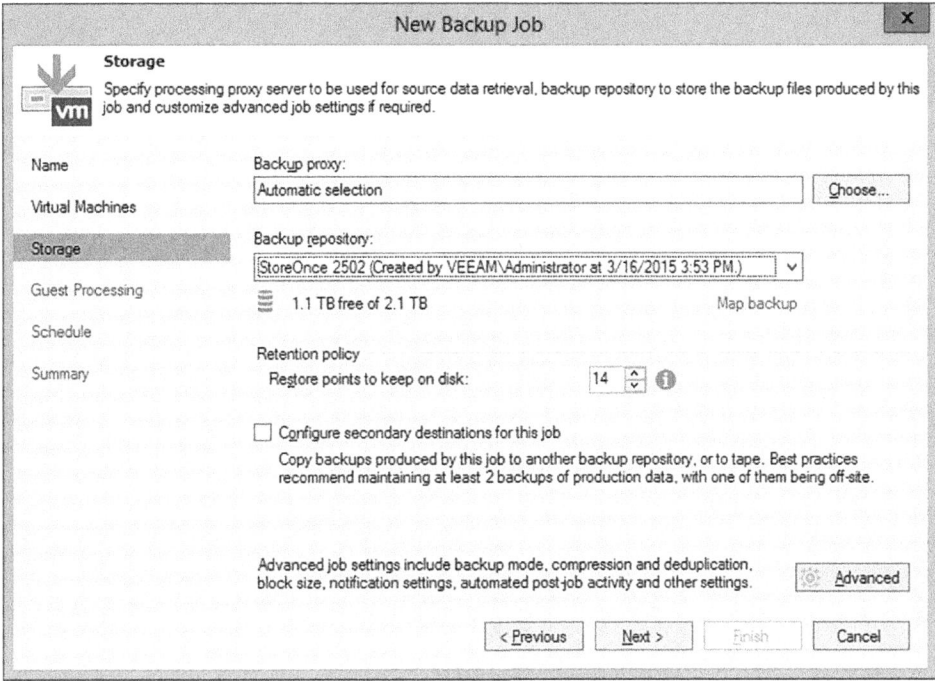

Figure 15-4 Example of the Veeam Interface

At this point, the installation is complete. The performance tests shown earlier were produced immediately after the installation.

Summary

The chapter covered a widely used technique of Veeam® software working with dedicated backup devices. This type of solution, with varying degrees of backup capacity, is mainstream and can be designed and implemented quickly. The sizing is a key factor in ensuring that the solution works for both the current capacity of the environment as well for growth.

16 Hyper Converged Backup and Recovery

INTRODUCTION

This solution covers the implementation of an HPE Hyper Converged 250 (HC250) with backup and recovery. This is a multi-tenant hosted environment, and some key aspects of the Virtual Machines (VMs) running on the HC250 are covered in the chapter as well. The HC250 has compute, storage, and networking resources to support a virtualized environment. The following is a list of goals for this solution:

- Reduce footprint and cost from traditional server, network, and storage by implementing the HC250
- Reduce operational expenses, reduce power and cooling, and simplify operations for local staff
- Implement virtual machines quickly on an appliance designed for this purpose
- Implement an effective backup and recovery solution to meet the backup window service-level agreements for longer retention time of data

Solution overview

The following is a list of the key components of the HC250, including both VMs running in the environment, plus the related backup and recovery solution:

- Four-server HC250 System running VMware vSphere
- Four-HPE HC250 software License to User (LTU) for VMware vSphere 6.0
- Eight-VMware vSphere Enterprise licenses for one processor
- Four-VMware vCenter Server Standard
- Two-10TB StoreVirtaul Virtual Storage Appliance (VSA) licenses for storage capacity expansion
- Two-HP DL380 Gen9 24 Small Form Factor (SFF) Configured to Order Server (CTO)
- One-HPE StoreOnce 3500 backup solution used to back up the environment
- HPE Data Protector backup software
- One-MSL 2024 tape library for off-site storage

CHAPTER 16
Hyper Converged Backup and Recovery

Figure 16-1 shows the HC250 and related backup components.

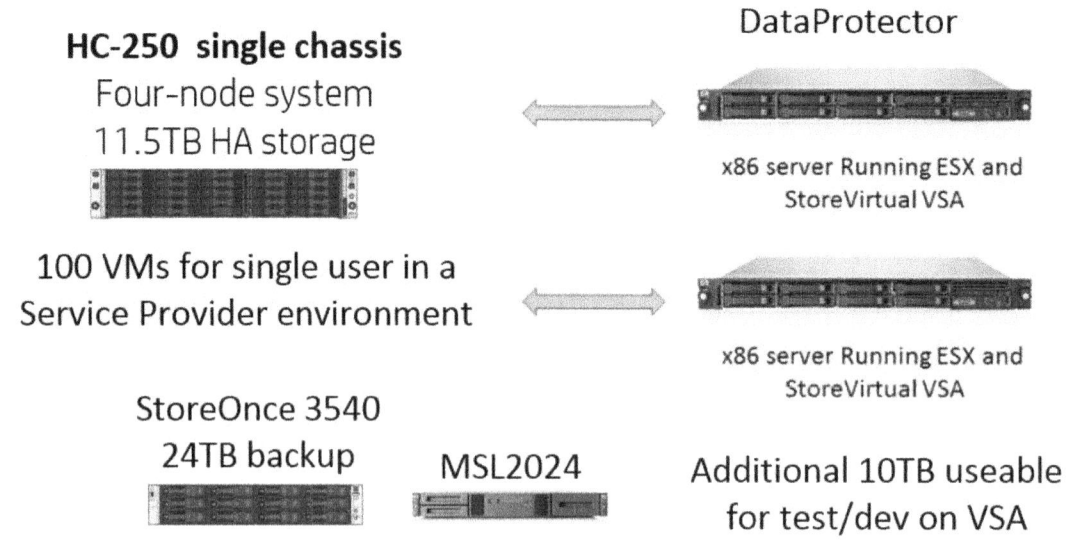

Figure 16-1 HC250 and backup solution

Note that the HPE Apollo 2U chassis has four independent servers in it. Figure 16-1 also shows the storage unit for the HC250 with backup software, backup disk, and backup tape for a complete end-to-end solution.

How did we arrive at this solution?

For the hosting aspect of the design, the performance per VM was the key parameter. There are two performance numbers related to data transfer used in this section:

- IOPs that are Input/Output (IO) operations per second. IOPs relate to random fetches of data and not sequential data movement. As you will see later in this chapter, we achieved 65 IOPs per virtual machine and had 25 virtual machines for a total of roughly 1625 IOPs. These random fetches of data will typically be much lower than sequential data movement of large blocks as described next.

- The second is MB/s that are mega-bytes per second. This is often referred to as "throughput" and is the number of MB transferred per second or, generally speaking, the maximum amount of data that you can transfer per second. You can imagine large blocks of data being transferred very quickly. The following is a formula that you can use to calculate IOPs and an example calculation:

MB/s = IOPs × KB per IO/1024

MB/s = 1650 × 512/1024 = 825 MB/s

This is an example calculation only to give you an idea of throughput in MB/s.

In our example in this chapter, we implemented the following HC250 solution with the parameters shown (Table 16-1):

Table 16-1 HC250 device characteristics

Hardware Features	Value
Rack footprint	2U per appliance
Number of compute nodes per appliance	4 nodes
Number of storage nodes per appliance	4 nodes
CPUs	96 Cores @ 2.6 GHz (24 per node) Intel ES-2680v3
Memory	2 TB DDR3 memory total (512 GB/node)
Networking ports	8 10 Gbe Ports (2 per node)

Testing was performed to ensure that the number of tested VMs did not exceed more than 70% of the process and memory utilization.

For this backup and recovery solution, an 8-hour window had to be achieved. Table 16-2 is a summary of the HPE StoreOnce device used in this design:

Table 16-2 StoreOnce device characteristics

	StoreOnce 3540
Form factor	2U scalable rack
Total capacity useable	15.5 TB
Total capacity used	8 TB compressed and de-duplicated
Backup performance	5 TB per hour
Restore performance	2.5 TB per hour
Max Fan-in/backup targets	24 targets

The upcoming sections have the detail of the hardware and software inventory of the design.

Hardware inventory

Table 16-3 shows the hardware Bill of Material (BOM) for the complete hardware in this solution that includes software not related specifically to backup:

CHAPTER 16
Hyper Converged Backup and Recovery

Table 16-3 BOM for HC250 and backup

Item#	Qty	Part#		HPE HC250 4-Node HyperConverged Appliance
				Description
1	1	M0T03B		HPE HC 250 System for VMware vSphere Sys
2	4	M0T04B		HPE Hyper Converged 250 Node
3	4	M0T04B	0D1	Factory integrated
4	4	793028-B21		HP XL1x0r Gen9 E5-2680v3 Kit
5	4	793028-B21	0D1	Factory integrated
6	4	793028-L21		HP XL1x0r Gen9 E5-2680v3 FIO Kit
7	16	728629-B21		HP 32GB 2Rx4 PC4-2133P-R Kit
8	16	728629-B21	0D1	Factory integrated
9	24	781518-B21		HP 1.2 TB 12 G SAS 10 K 2.5 inches SC ENT HDD
10	24	781518-B21	0D1	Factory integrated
11	4	665243-B21		HP Ethernet 10Gb 2P 560FLR-SFP+ Adptr
12	4	665243-B21	0D1	Factory integrated
13	4	P9B51A		HPE HC 250 SW LTU for VMware vSphere 6.0
14	4	P9B51A	0D1	Factory integrated
15	1	676277-B21		HP 36pin Suv Dongle Cord Kit
16	1	676277-B21	0D1	Factory integrated
17	2	720620-B21		HP 1400W FS Plat Pl Ht Plg Pwr Spply Kit
18	2	720620-B21	0D1	Factory integrated
19	1	H1K92A3		HPE 3Y Proactive Care 24x7 Service
20	1	H1K92A3	YMW	HP CS 250-HC StoreVirtual System Support
21	4	H1K92A3	YMX	HP CS 250-HC StoreVirtual Node Support
22	4	H1K92A3	YMY	HP CS 250-HC StoreVirtual SW LTU Support
23	1	HA114A1		HP Installation and Startup Service
24	1	HA114A1	5WG	HPE 200 s HC StoreVirtual Startup SVC
25	8	BD713A		VMw vSphere Ent 1P 3 yr SW
26	4	BD725A		VMw vCntr Srv Std 3 yr SW
27	1	P8A90AAE		HP SV VSA 2014 4TB 3pk 3 yr DPS E-LTU
28	1	H1K92A3		HPE 3Y Proactive Care 24x7 Service
29	8	H1K92A3	R5H	VMw vSphere Ent 1P 3 yr SW
30	4	H1K92A3	R61	VMw vCntr Srv Std 3 yr SW
31	1	H1K92A3	YN9	HPE SV VSA 2014 4TB 3pk 3 yr LTU Support

Table 16-3 Continued.

Item#	Qty	Part#		HPE HC250 4-Node HyperConverged Appliance Description
				HyperConverged Storage Expansion Node
32	1	767032-B21		HP DL380 Gen9 24SFF CTO Server
33	1	767032-B21	ABA	US-English localization
34	1	719053-L21		HP DL380 Gen9 E5-2603v3 FIO Kit
35	2	726718-B21		HP 8GB 1Rx4 PC4-2133P-R Kit
36	2	726718-B21	0D1	Factory integrated
37	13	781518-B21		HP 1.2TB 12G SAS 10K 2.5 inches SC ENT HDD
38	13	781518-B21	0D1	Factory integrated
39	1	665243-B21		HP Ethernet 10Gb 2P 560FLR-SFP+ Adptr
40	1	665243-B21	0D1	Factory integrated
41	1	749974-B21		HP Smart Array P440ar/2G FIO Controller
42	1	727250-B21		HP 12Gb DL380 Gen9 SAS Expander Card
43	1	727250-B21	0D1	Factory integrated
44	1	733660-B21		HP 2U SFF Easy Install Rail Kit
45	1	733660-B21	0D1	Factory integrated
46	1	700139-B21		HP 32GBmicroSDMainstream Flash Media Kit
47	1	700139-B21	0D1	Factory integrated
48	2	720478-B21		HP 500W FS Plat Ht Plg Pwr Supply Kit
49	2	720478-B21	0D1	Factory integrated
50	1	P8B31A		HP OV w/o iLO 3 yr 24x7 FIO Phys 1 LTU
51	1	BD505A		HP iLO Adv incl 3 yr TS U 1-Svr Lic
52	1	BD505A	0D1	Factory integrated
53	1	H1K92A3		HPE 3Y Proactive Care 24x7 Service
54	1	H1K92A3	R2M	HPE iLO Advanced Non Blade -3 yr Support
55	1	H1K92A3	SVP	HP One View w/o Ilo Supp
56	1	H1K92A3	TT3	HPE ProLiant DL380 Gen9 Support
57	1	HA114A1		HP Installation and Startup Service
58	1	HA114A1	5A6	HPE Startup 300 Series OS SVC
59	1	E8H73AAE		VMw vSph Std 1P 3 yr Channel E-LTU
60	1	TC486AAE		HP SV VSA 2014 10TB 3 yr E-LTU
61	1	H1K92A3		HPE 3Y Proactive Care 24x7 Service
62	1	H1K92A3	R5E	VMw vSphere Std 1P 3 yr SW
63	1	H1K92A3	RWR	HPE SV VSA 2014 10TB 3 yr Support

CHAPTER 16
Hyper Converged Backup and Recovery

Table 16-3 Continued.

Item#	Qty	Part#		HPE HC250 4-Node HyperConverged Appliance
				Description
				HPE StoreOnce Backup Appliance
64	1	BB914A		HPE StoreOnce 3540 24TB System
65	1	BB926A		HPE StoreOnce 10GbE Network Card
66	1	BB926A	0D1	Factory integrated
67	1	BB949A		HPE StoreOnce 10GbE Netwrk Exp LTU
68	1	BB949A	0D1	Factory integrated
69	1	H1K92A3		HPE 3Y Proactive Care 24x7 Service
70	1	H1K92A3	XDQ	HPE StoreOnce 3500 24TB Backup Supp
71	1	HA124A1		HP Technical Installation Startup SVC
72	1	HA124A1	55Q	HPE StoreOnce System startup SVC
				HPE DL360 Backup Server—MSL2024 Tape Library
84	1	755259-B21		HP DL360 Gen9 4LFF CTO Server
85	1	755259-B21	ABA	US-English localization
86	1	755374-L21		HP DL360 Gen9 E5-2603v3 FIO Kit
87	2	726718-B21		HP 8GB 1Rx4 PC4-2133P-R Kit
88	2	726718-B21	0D1	Factory integrated
89	2	804680-B21		HP 1.2TB 6G SATA WI-2 LFF SCC SSD
90	2	804680-B21	0D1	Factory integrated
91	1	766211-B21		HP DL360 Gen9 LFF P440ar/H240ar SAS Cbl
92	1	766211-B21	0D1	Factory integrated
93	1	789388-B21		HP 1U LFF Gen9 Mod Easy Install Rail Kit
94	1	789388-B21	0D1	Factory integrated
95	1	665243-B21		HP Ethernet 10Gb 2P 560FLR-SFP+ Adptr
96	1	665243-B21	0D1	Factory integrated
97	1	749974-B21		HP Smart Array P440ar/2G FIO Controller
98	1	726911-B21		HP H241 Smart HBA
99	1	726911-B21	0D1	Factory integrated
100	2	720478-B21		HP 500W FS Plat Ht Plg Pwr Supply Kit
101	2	720478-B21	0D1	Factory integrated
102	1	P8B31A		HP OV w/o iLO 3 yr 24x7 FIO Phys 1 LTU
103	1	339778-B21		HP Raid 1 Drive 1 FIO Setting
104	1	BD505A		HP iLO Adv incl 3 yr TS U 1-Svr Lic

Table 16-3 Continued.

Item#	Qty	Part#		HPE HC250 4-Node HyperConverged Appliance
				Description
105	1	BD505A	0D1	Factory integrated
106	1	H1K92A3		HPE 3Y Proactive Care 24x7 Service
107	1	H1K92A3	R2M	HPE iLO Advanced Non Blade - 3 yr Support
108	1	H1K92A3	SVP	HP One View w/o Ilo Supp
109	1	H1K92A3	TT5	HPE ProLiant DL360 Gen9 Support
110	1	AK379A		HP MSL2024 0-Drive Tape Library
111	1	H1K92A3		HPE 3Y Proactive Care 24x7 Service
112	1	H1K92A3	80N	MSL2024 Library Support
113	2	N7P37A		HPE MSL LTO-7 SAS Drive Upgrade Kit
114	2	C7978A		HP Ultrium Universal Cleaning Cartridge
115	1	C7977AN		HPE LTO-7 Ultrium Non Custom Lbl 20 Pk

Software inventory

Table 16-4 shows the software components that are implemented for this backup and recovery solution:

Table 16-4 Backup-related software used to implement this solution

Item#	Qty	Part#		HPE DataProtector SW
				Description
73	1	B6961CAE		HP Data Prot Stater Pack for Linux E-LTU
74	1	B6963AAE		HP DP drive extn WIN/Netware/Linux E-LTU
75	4	B6965BAE		HP DP On-line Backup for Windows E-LTU
76	2	B7038BAE		HP DP Advanced Backup to Disk 10TB E-LTU
77	1	TD586EAE		HP Data Protector 9.00 Eng SW E-Media
78	1	H7J34A3		HPE 3Y Foundation Care 24x7 Service
79	1	H7J34A3	1QK	HP Software 1QK Supp
80	4	H7J34A3	1QL	HP Software 1QL Supp
81	1	H7J34A3	43B	HP Software 43B Supp
82	2	H7J34A3	7S5	HP Software 7S5 Supp
83	1	H7J34A3	97D	HP Software 97D Supp

CHAPTER 16
Hyper Converged Backup and Recovery

Solution validation and success criteria

The final design, both HC250 and backup and recovery, was crafted as a result of the following activities:

- On-site assessment of capacity and performance needed for production hosting of 100 virtual machines.
- The ability to provision virtual machines for different workloads in 30 minutes.
- The need for a test/dev environment with the ability to move from production to test/dev with near online data.
- Backup and recovery to meet service-level agreements of under 8 hours.

These four areas were needed to ensure that we met the success criteria.

Implementation, deployment, and testing

Initial installation and deployment of complete solution were done within one day, and day two was set aside for a series of tests to meet success criteria that include the following:

- Replication of workload consisting of 25 of the top critical virtual machines to HC250.
- Movement of live data via snapshot replication to test/dev environment with current data no less than 3 hours old.
- Provisioning of different virtual machine workloads in 30 minutes.
- Backup of the complete environment of approximately 5 TB over 10 Gb Ethernet to meet backup and recovery service-level agreements.

Test workload results

Several tests were run to ensure that the design and implementation met both ability to scale to the 100 Virtual Machines and the backup and recovery requirements for the environment including the ones listed below:

- The 25 virtual machines tested at an average of 65 IOPs each that is roughly 1625 IOPs at 95% of I/O capacity, which is close to the maximum that can be expected.
- Used internal HC250 StoreVirtaul snapshot replication to test/dev at 60 minutes intervals: one set of 25 snapshots per 24-hour period.
- Used default virtual machine provision templates of small, medium, and large workloads with an average time to complete in 5 minutes.
- Backed up 5 TBs of data using DataProtector software to the StoreOnce 3500 in one hour with a restore time just under two.

These are the key tests identified by the system administrator of this multi-tenant environment. There may be additional tests that you want to run, such as replication, according to the specific needs of your environment.

Summary

This chapter focused on the backup and recovery of HC250 and touched on some key aspects of provisioning and running VMs on HC250. The old environment of separate server, storage, and networking was replaced with the HC250 taking into account capacity, performance, growth, and other factors such as the ease of provisioning. The HC250, with its ease-of-provisioning, turned out to be the ideal platform for this hosted, multi-tenant environment.

The backup and recovery of the environment is very important in that the tenants of this environment expect the backup and recovery requirements to be met. In this example, a solution that can be easily expanded was used to accommodate current storage capacity, plus modest additional capacity as well as the desired performance to meet backup service-level requirements. Extensive testing was performed to ensure that these requirements were met so that the solution could be deployed with confidence in backup and recovery.

The production environment covered in this chapter represented roughly 25% of the multi-tenant workload. There was enough capacity in the HC250 solution, as well as backup and recovery capacity, to move the remaining 75% of the workload to the HC250. All the performance requirements were met when the remainder of the environment was moved over, including provisioning.

The backup and restore performance requirements were also met when the additional 75% of the workload was moved as well. Backup of 11.5 TB was complete within the prescribed window.

17 Disaster Recovery

INTRODUCTION

This chapter covers Disaster Recovery (DR) for an installation that has been successfully running two active-active production and DR data centers. The chapter focuses primarily on the storage aspects of crafting the DR solution because most of the DR design involved storage.

The firm is now separating into two companies that will operate independently of one another. Both the physical data structure, as well as the data on the servers, would need to be modified to fit the new business model.

Although the two new companies would no longer be able to utilize each other's physical space for their DR needs, and a new location for the DR was imperative, it was possible that each company could take ownership and reuse components of the existing infrastructure.

A solution was crafted that enables both businesses to maintain production and DR processes, while adhering to the following criteria:

- Minimize disruption of the current DR process: specifically, the amount of time during which there is a switch between DR targets
- Minimize the impact on IT staff and operations
- Minimize disruption of the business function while bifurcating data from both production servers and reconfiguring business applications as appropriate
- Maintain current, or higher, performance levels
- Determine alternate location of DR servers for each company
- Minimize cost because both firms would be required to procure two sets of the technical solution
- There is a tight timeframe during which this transition can be performed

The next section covers the current and new configurations for this solution.

CHAPTER 17
Disaster Recovery

Current and new configurations

The current configuration consisted of the following components:

- Each site contained the following infrastructure components:
    - 3PAR Storage array for primary storage
    - EVA Storage array for secondary storage
    - StoreOnce deduplication backup appliance
    - Multiple C7000 Blade Chassis containing both HP-UX and ESX blades (G5/G6)
    - ProLiant DL server-based Exchange environment

Figure 17-1 depicts the configuration of the existing infrastructure.

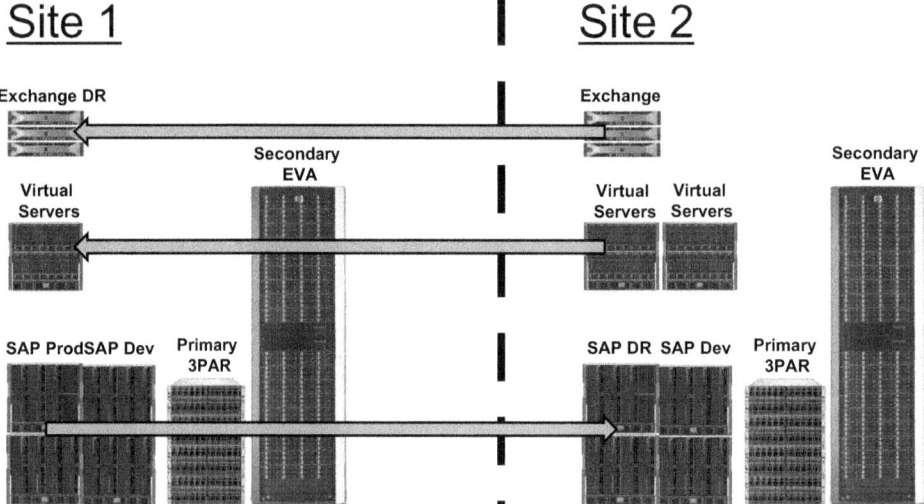

Figure 17-1 Existing configuration of Sites 1 and 2

Figure 17-2 depicts the new configuration.

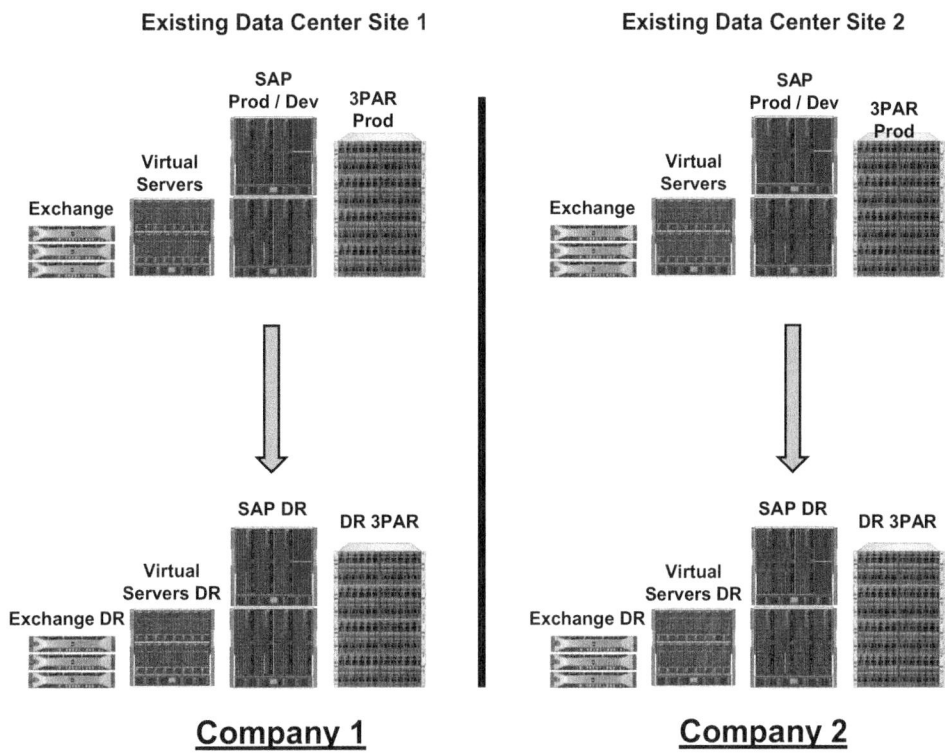

Figure 17-2 New solution

Note that in this new solution there are two sites for each of the two firms with no shared components.

The next section covers the new solution overview.

The new solution overview

Workloads, capacities, and future growth expectations for both companies were evaluated and considered for the new design. The goal of minimizing expense as well as leveraging all of the existing infrastructure were the basis for the design. The following is a summary of the new solution.

CHAPTER 17
Disaster Recovery

Storage:

- Two new 3PAR 7440 Hybrid Arrays used for DR. Each array consisted of:
    - 42u Storage centric rack
    - Four controller nodes
    - Four 16 Gb quad port FC adapters
    - 12 SFF drive shelves
    - 4 LFF drive shelves
    - 16 1.92 TB SSD drives
    - 192 1.8 TB 10k drives
    - 24 4 TB 7.2k drives
    - Replication software suite
    - Data Optimization software suite
    - Application Integration software suite
- Two new StoreOnce 4700 De-Dup Appliances
    - Catalyst Integration
- Two new SN6000B 16Gb Fibre Channel switches each site

Servers:

- Company 1 Production site
    - Reuse two existing c7000 chassis with 15 new gen 9 blades
- Company 1 DR site
    - Reuse two existing c7000 chassis with 15 new gen 9 blades
    - One new c7000 chassis with 16 gen 9 blades
- Company 2 Production site
    - Reuse two existing c7000 chassis with 15 new gen 9 blades
- Company 2 DR site
    - Reuse one existing c7000 chassis and one new c7000 chassis with 15 gen 9 blades
    - One new c7000 chassis with 16 gen 9 blades

Figure 17-3 depicts the racks in the new solution. Each entity will be supplied with an identical storage solution to be installed in separate cages within the HPES colocation facility:

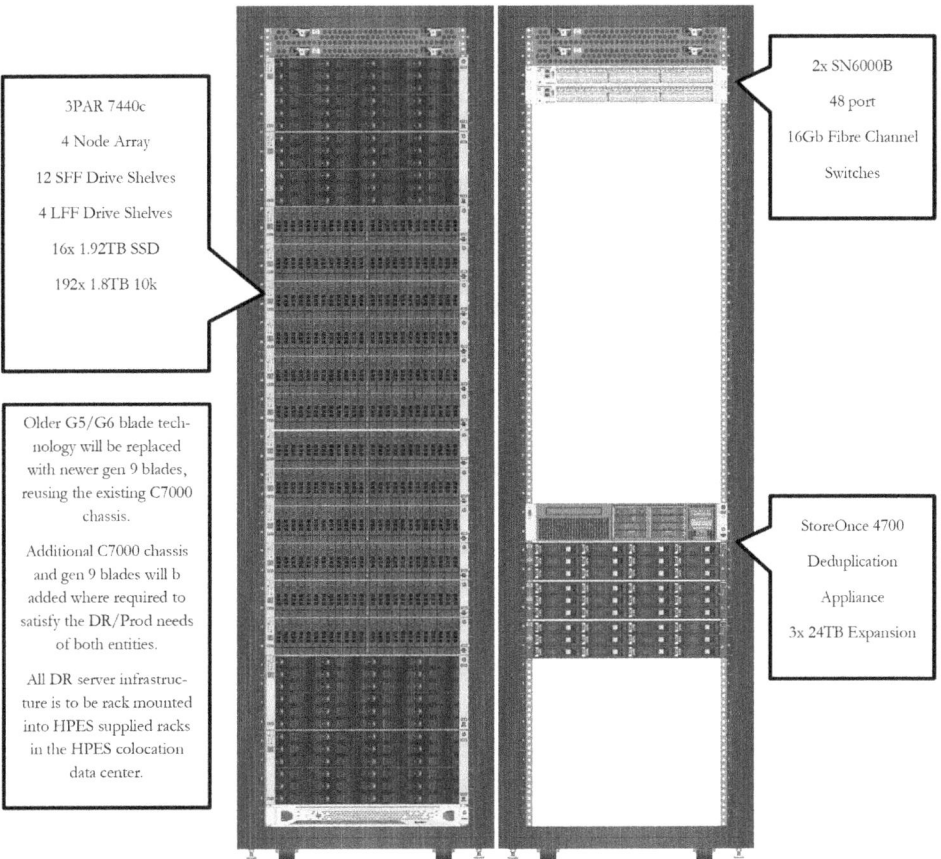

Figure 17-3 New solution rack diagram

How did we arrive at this solution?

This section focuses on the storage considerations of crafting the DR solution. The first goal of the storage solution was to ensure that we could satisfy the existing capacity and performance requirements as well as the projected future growth. One of the objectives was to build a configuration that would allow both production and DR to operate simultaneously with little impact on users and applications.

The existing solution could not accommodate production and DR workloads from both sites simultaneously; therefore, we could not use it as a reference design moving forward. We analyzed

CHAPTER 17
Disaster Recovery

performance and capacity utilization over several periods to determine the benchmark for the design. In any storage design, there is always a compromise between the number and type of spindles and the resulting usable capacity and performance potential.

We decided to reduce the number of disk tiers to three from four, as well as increase the overall Solid State Disk (SSD) capacity percentage for increased Adaptive Optimization effect. To optimize the initial costs, fewer spindles were configured with the understanding that adjustments may be needed in the future to keep pace with performance growth.

The new DR target arrays were configured with approximately 30% increase in capacity to allow for growth over the next 18 months. Ample free slots are available in the near-line tier for additional bulk storage.

The existing configuration layouts and specifications for Site 1 and Site 2 are shown in the illustrations of Figures 17-4 and 17-5.

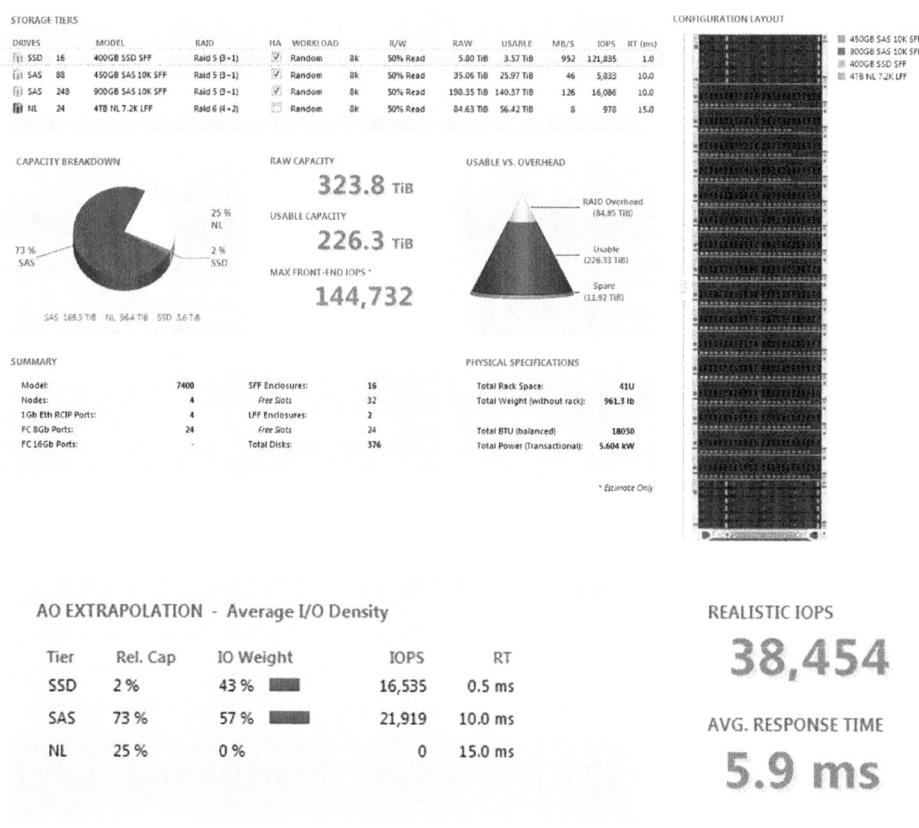

Figure 17-4 Existing solution Site 1

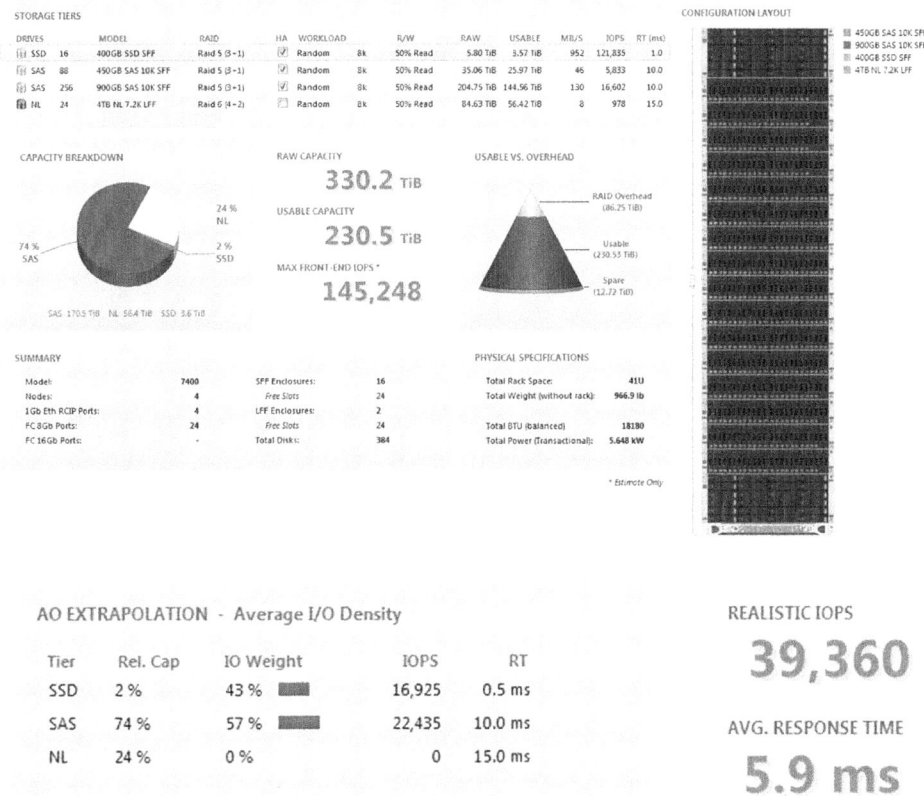

Figure 17-5 Existing solution Site 2

CHAPTER 17
Disaster Recovery

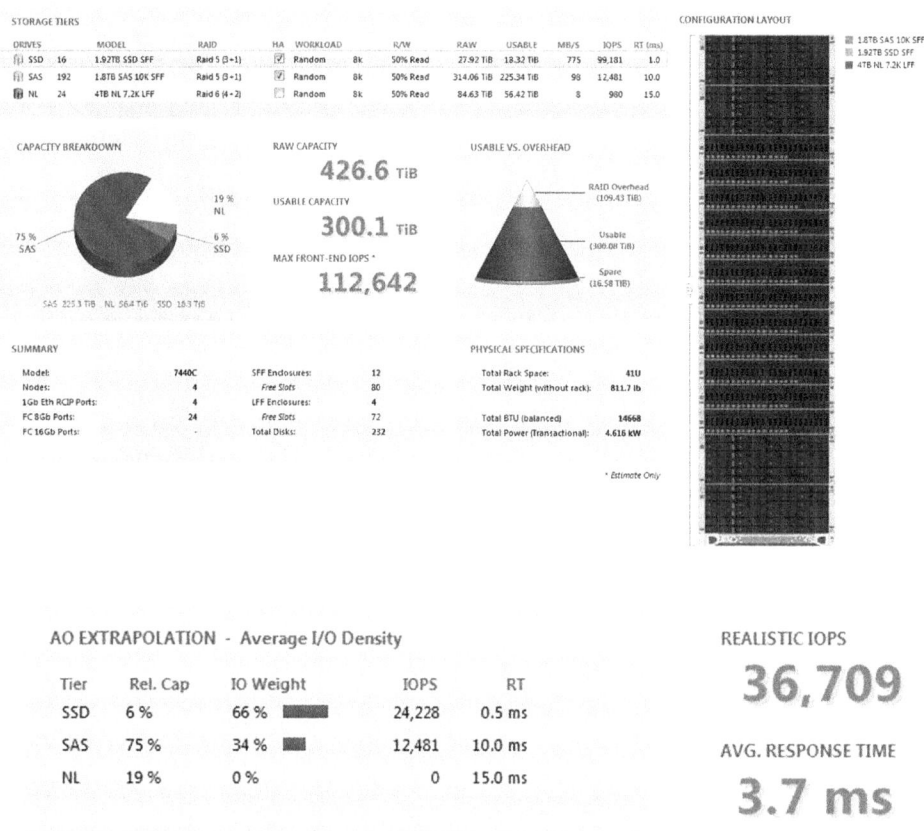

Figure 17-6 New DR targets (Two identical solutions)

As you can see from Figures 17-4 to 17-6, using larger capacity disk drives (1.8 TB vs 900 GB) allowed us to increase the usable capacity by 30%, while maintaining the same physical footprint and similar performance characteristics.

Hardware and software inventory

Table 17-1 shows the Bill of Materials (BOM) for this DR solution.

Several optional 3PAR software features were used in the solution. Data protection capabilities such as Virtual Copy and Replication are central to the function of this solution, whereas autonomic features such as Adaptive Optimization and Dynamic Optimization help to reduce the overall cost and improve operational efficiency.

Table 17-1 Bill of Materials

Line No.	Qty	Part Number		Description
HPE 3PAR StoreServ Array					
				Hardware
3	1	E7X84A		HP 3PAR STORESERV 7440C 4N ST CENT BASE
5	4	E7X47A		HP 3PAR 7000 2-PT 16GB FC ADAPTER
29	10	QR490A		HP M6710 2.5IN 2U SAS DRIVE ENCLOSURE
59	4	QR491A		HP M6720 3.5IN 4U SAS DRIVE ENCLOSURE
7	16	E7Y57A		HP M6710 1.92TB 6G SAS 2.5IN CMLC SSD
9	192	K0F26A		HP M6710 1.8TB 6G SAS 10K 2.5IN HDD
61	24	H6Z87A		HP M6720 4TB 6G SAS 7.2K 3.5IN HDD
80	2	AP879A		HP 6M EXPANSION CABLE KIT
45	8	QK734A		HP PREMIER FLEX LC/LC OM4 2F 5M CBL
				Software
11	1	BD374A		HP 3PAR 7440C OS SUITE BASE LTU
13	232	BD381A		HP 3PAR 7440C OS SUITE DRIVE LTU
15	1	BD382A		HP 3PAR 7440C REPLICATION SUITE BASE LTU
17	232	BD383A		HP 3PAR 7440C REPLICATION STE DRIVE LTU
19	1	BD404A		HP 3PAR 7440C DATA OPT ST V2 BASE LTU
21	232	BD405A		HP 3PAR 7440C DATA OPT ST V2 DRIVE LTU
23	1	BD375A		HP 3PAR 7440C REPORTING SUITE LTU
25	1	BD376A		HP 3PAR 7440C APP SUITE VMWARE LTU
27	1	BD378A		HP 3PAR 7440C APP SUITE SQL LTU
74	1	BD362AAE		HP 3PAR STORESERV MGMT/CORE SW E-MEDIA
75	1	BD363AAE		HP 3PAR OS SUITE E-MEDIA
76	1	BD371AAE		HP 3PAR APP SUITE FOR SQL E-MEDIA
77	1	BD372AAE		HP 3PAR APP SUITE FOR VMWARE E-MEDIA
78	1	BD373AAE		HP 3PAR REPORTING SUITE E-MEDIA
				Support and Services
83	1	H1K92A5		HP 5Y 4 HR 24X7 PROACTIVE CARE SVC
89	1	H1K92A5	RZ6	HP 3PAR 7440C/50OS SUITE BASE LTU SWSUPP
90	1	H1K92A5	RZ7	HP 3PAR7440C/50REPLSUITE BASE LTU SWSUPP
91	1	H1K92A5	RZJ	HP 3PAR7440C/50REPORTINGSUITE LTU SWSUPP
92	2	H1K92A5	RZK	HP 3PAR 7440C/50 AP SUITE LTU SW SUPP

CHAPTER 17
Disaster Recovery

Table 17-1 Continued.

		HPE 3PAR StoreServ Array		
Line No.	Qty	Part Number		Description
93	232	H1K92A5	S7Y	HP 3PAR7440C/50OS SUITE DRIVE LTU SWSUPP
94	232	H1K92A5	S7Z	HP 3PAR7440C/50REPLSUITEDRIVE LTU SWSUPP
95	1	H1K92A5	SDT	HP 3PAR7440C/50DATAOPTSTV2BASELTU SWSUPP
96	232	H1K92A5	SDU	HP 3PAR7440C/50DATAOPTSTV2DRV LTU SWSUPP
97	16	H1K92A5	TQV	HP 3PAR 7000 1.92TB SAS CMLC SSD HW SUPP
98	1	H1K92A5	TRJ	HP 3PARSTORESERV7440C/50C 4N BASE HWSUPP
99	4	H1K92A5	TRM	HP 3PAR 7000 2-PT 16GB FC ADAPTER HWSUPP
101	216	H1K92A5	WUT	HP 3PAR 7000 DRIVES OVER 1TB SUPPORT
102	14	H1K92A5	WUW	HP 3PAR 7000 DRIVE ENCLOSURE SUPPORT
106	1	HA124A1		HP TECHNICAL INSTALLATION STARTUP SVC

		HPE SAN		
Line No.	Qty	Part Number		Description
		Hardware		
55	2	QR481B		HP SN6000B 16GB 48/48 PWR PCK+ FC SWITCH
57	96	QK724A		HP B-SERIES 16GB SFP+SW XCVR
65	96	QK735A		HP PREMIER FLEX LC/LC OM4 2F 15M CBL
		Software		
79	1	TC472AAE		HP INTELLIGENT INFT ANALYZER SW V2 E-LTU
		Support and Services		
83	1	H1K92A5		HP 5Y 4 HR 24X7 PROACTIVE CARE SVC
86	2	H1K92A5	QAL	HP SN6000B 16GB 48/24 PPCK+ FC JW SUPP
88	1	H1K92A5	RWG	HP INTELLIGENT INFRA ANALYZER LTU SUP
106	1	HA124A1		HP TECHNICAL INSTALLATION STARTUP SVC

		HPE StoreOnce Deduplication Appliance	
Line No.	Qty	Part Number	Description
		Hardware	
81	1	BB879A	HP STOREONCE 4700 24TB BACKUP
82	2	BB881A	HP STOREONCE 4500/4700 24TB UPGRADE KIT
		Software	
73	1	BB889AAE	HP STOREONCE 4400/4700 CATALYST E-LTU

Table 17-1 Continued.

HPE StoreOnce Deduplication Appliance			
Line No.	Qty	Part Number	Description
			Support and Services
83	1	H1K92A5	HP 5Y 4 HR 24X7 PROACTIVE CARE SVC
84	1	H1K92A5 27Z	HP STOREONCE 43/47 BACKUP SYSTEM HW SUPP
85	2	H1K92A5 28A	HP STOREONCE43/45/4700 CAP UPG KIT SUPP
87	1	H1K92A5 QBT	HP CATALYST 4400 LTU SW SUPP
106	1	HA124A1	HP TECHNICAL INSTALLATION STARTUP SVC
107	1	HA124A1 5TY	HP STOREONCE CATALYST LVL1 SOLUTION SVC

HPE Racks			
Line No.	Qty	Part Number	Description
1	1	BW904A	HP 642 1075MM SHOCK INTELLIGENT RACK
2	1	BW904A 001	HP FACTORY EXPRESS BASE RACKING SERVICE
43	1	TK808A	HP RACK FRONT DOOR COVER KIT
47	4	H5M58A	HP 4.9KVA 208V 20OUT NA/JP BPDU
49	1	BW932A	HP 600MM RACK STABILIZER KIT
51	1	BW906A	HP 42U 1075MM SIDE PANEL KIT
53	1	BW904A	HP 642 1075MM SHOCK INTELLIGENT RACK
54	1	BW904A 001	HP FACTORY EXPRESS BASE RACKING SERVICE
63	1	TK808A	HP RACK FRONT DOOR COVER KIT
67	4	H5M58A	HP 4.9KVA 208V 20OUT NA/JP BPDU
69	1	BW932A	HP 600MM RACK STABILIZER KIT
71	1	BW906A	HP 42U 1075MM SIDE PANEL KIT
103	1	BW906A	HP 42U 1075MM SIDE PANEL KIT
104	1	BW932A	HP 600MM RACK STABILIZER KIT
105	1	TK808A	HP RACK FRONT DOOR COVER KIT

Solution validation and success criteria

The final design was crafted as a result of on-site testing that was performed to ensure that we met the success criteria.

Deployment overview

Unraveling the existing mixed infrastructure and data from the combined company into separate entities poses logistical challenges that require careful planning and coordinated execution.

One specific challenge is the "shell game" with production data and available space. The existing arrays do not have enough free space in order to maintain a production copy as well as DR copy of the data at each location. Therefore, it is important to stand-up the new DR arrays quickly in order to reconfigure the DR targets from the existing arrays to the new DR arrays located at the HP colocation facility.

Repointing the DR copies to the new arrays will free space on the existing arrays for copies of the production data at each site. Once each primary site has a copy of the production data, a final reconfiguration of the DR targets to the appropriate DR array can take place.

This shell game will take place in a phased approach as to minimize the performance impact of data movement as well as to ensure that DR protection is maintained within an acceptable window.

A decision was made to start each company with a full copy of the production dataset, even though each entity only requires a subset of the whole. This method was chosen to help minimize the amount of time and the complexity of bifurcating the data beforehand. By performing the data cleansing post-split, each company can more carefully assess the needs of their business and the value of the various datasets to them. The risks to the schedule, as well as the risk to the datasets, are also greatly reduced.

Validation

The solution itself does not require extensive validation as the new architecture essentially is the same as the existing in terms of components, management tools, and process flow. The risk and validation is more with the methodologies employed to bifurcate the replication relationships and shuffle the data. Since this is planned for a phased approach, any lessons learned in the initial phases can be applied as the project progresses.

Education and guidance will be key to success as the employees and skills sets are transitioned along with the infrastructure and business units. Each new entity will have skill gaps as employees are assigned to one business unit or another. We worked closely with one of our leading partners who helped deliver these services. They provided onsite resources for training and transitioning.

Training, consulting services, and general guidance are planned to assist the entities with a smooth transition in managing the ongoing operations of their respective environments.

Summary

To accommodate a company separation across two sites, this DR solution was designed with the goal of minimizing expense while leveraging the existing data center infrastructure. Opportunities for optimizing Workloads, extending storage capacities, and meeting the future growth expectations for both companies were evaluated and considered for the new design. A fresh objective was to build a configuration that would allow both production and DR to operate simultaneously with little impact on users and applications.

Despite logistical challenges, several benefits were realized in the solution design. The use of larger capacity disk drives increased usable capacity by 30%, while maintaining the same physical footprint and performance. Optional 3PAR software features such as Virtual Copy and Replication, and autonomic features such as Adaptive Optimization and Dynamic Optimization, were instrumental in reducing overall cost and improving operational efficiency.

Although the new architectures for both sites were similar in terms of components, management tools, and process flow, there were additional risks that had to be recognized due to organizational changes. To ensure for future success, the consulting services for this solution included additional considerations such as coverage of skill gaps. Onsite resources for training have been planned to enable a smooth transition in managing the ongoing operations of each environment.

18 An Asymmetric Approach to Hadoop

INTRODUCTION

This chapter describes a Hadoop deployment that breaks away from the traditional "symmetric" approach where compute and storage reside on the same node. Using HPE workload and density-optimized (WDO) servers, this solution creates an asymmetric cluster, so you can flexibly grow Hadoop's compute and storage needs independently in the future. This particular WDO solution covers the move from a traditional symmetric balanced and density optimized (BDO) Hadoop storage and server environment to a more flexible, space-efficient, and performant Hadoop environment.

Additional background on symmetric BDO and asymmetric WDO architectures on various Hadoop distributions is found at the following URL:

http://www8.hp.com/us/en/products/servers/high-performance-computing/hadoop.html

Design Factors

When designing a Hadoop environment in the past, the method of using traditional 2U-x86 servers with CPU and Storage has been the only option because most all vendors provided only this type of server packaging offering the only known, yet simple, building-block approach that was easy to expand. In many Hadoop environments, however, some design factors require a new approach to architecting Hadoop solutions, and they are described as follows:

- **Growth**—No two customers run Hadoop in exactly the same way or for the same purpose. We have noted that some customers require more disk capacity, while other customers require more compute power. Even the same customer seems to require more compute capacity today and then more disk capacity tomorrow. The WDO solution surpasses the traditional Hadoop (BDO) requirement of adding compute *and* storage at the same time by allowing compute and/or storage nodes to be added (scaled) independently. Therefore, a WDO customer can add what resource they need, when they need it; all while still allowing the similar building block approach of adding a single node at a time.

- **Performance**—Within any data-dense Hadoop architecture, performance is crucial because processing the data quickly allows organizations to rapidly analyze and gain actionable insight.

Hadoop ecosystem projects such as Apache SPARK, Apache Drill, and Impala are becoming more prevalent among customers to provide this faster "analysis into insight." SPARK, Drill, Impala, and similar codes' execution profiles typically require more dedicated Compute and RAM scale-out on compute nodes. Hadoop customers also tend to run more than one type of analysis code concurrently on their Hadoop cluster because the architecture of each analysis code may be better suited for one type of analysis more so than another. A classic example might be the use of SPARK for SQL analysis and the use of "R" for math and statistics—both operating on the same HDFS datasets. Since HPE Big Data Engineering helped create YARN container "labels," this allowed creation of WDO architectures where SPARK can run (dedicated) on some compute nodes, while "R" codes run (dedicated) on different compute nodes all in the same cluster and accessing all of the same data. Because SPARK and R can run dedicated on their own nodes, they get more RAM caching advantages that help get query answers faster. This is especially true compared with SPARK and R running together on the same server in traditional BDO Hadoop architectures. Most Hadoop customers are running Hadoop distros that include the YARN scheduler making all of this possible.

- **Flexibility**—HPE does not just build a 2RU rackmount server and actually innovates by building special purpose servers that are either compute or storage centric in their respective design centers. HPE have produced compute-centric servers (Apollo 2000) that remove several disk slots in favor of adding more CPU and RAM slots in a very dense 2U packaging. Complementary to Apollo 2000 compute nodes, HPE has designed the Apollo 4200 and 4500 to be extreme storage servers that maximize disk slots and minimize server CPU. Joining these two diverse server types using HPE 40 G network switches is what make WDO Hadoop possible and scale better than traditional Hadoop (BDO) Architectures.

- **Floor space**—Floor space cost is at a premium for many existing organizations or new data centers as many companies struggle to keep their footprint to a minimum. In traditional Hadoop deployments, a 2U general-purpose server is used, scaling compute and storage on the same server regardless of your compute or data storage needs. An asymmetric solution employs workload-focused, rack optimized Apollo servers that support either compute or storage. Compute-dense servers are used for compute and storage-dense servers for storage which allows the Hadoop cluster to increase either its storage capacity or compute capacity with specifically designed HPE servers required for the growth. The resulting rack-unit savings of asymmetric (WDO) Hadoop versus the traditional symmetric (BDO) Hadoop implementations is indeed compelling. In this case, the WDO solution requires only 17U for a full Hadoop cluster and switching infrastructure that leaves more than half a rack for cluster growth. The BDO traditional footprint is 40% more. This shows that WDO will use rack space more slowly as you scale out.

- **Power**—Because the asymmetric WDO Hadoop takes up less rack space, so it follows that it requires less power than the traditional BDO Hadoop architectures. Less power means less heat is generated, which means less cooling is required. This makes WDO a greener approach to BigData computing.

- **Simple deployment**—In a complex cluster like this, it is critical to simplify deployment, relying on HPE Factory Express services to package this as a complete solution. These limit delays in deployment once the equipment is on-site because the solution arrives with operating system installed, management tools installed, and networking configured. Using HPE's Cluster Management Utility (CMU) also allows for a simplified deployment and continued monitoring and management of the cluster.

- **Cost**—By separating the compute and storage you can focus on what needs to be added, either processing power or storage, without adding unnecessary and, for this reason, underutilized resources. As shown earlier, this efficient growth saves costs, floor space, power, and so on. It should also be noted that HPE BigData Engineering has tested WDO versus BDO Hadoop using all HPE gear. For this comparison, care was taken to use the same list price cost for each WDO and BDO Hadoop configuration and then run the same Hadoop benchmark codes on each cluster. Overall, WDO ended up getting faster answers with less compute servers and less footprint. This surely supports the initial assertion above that underutilized resources are minimized using WDO Asymmetric Hadoop configurations.

Solution overview

The overall environment of Hadoop compute and storage nodes in the cluster are depicted in Figure 18-1.

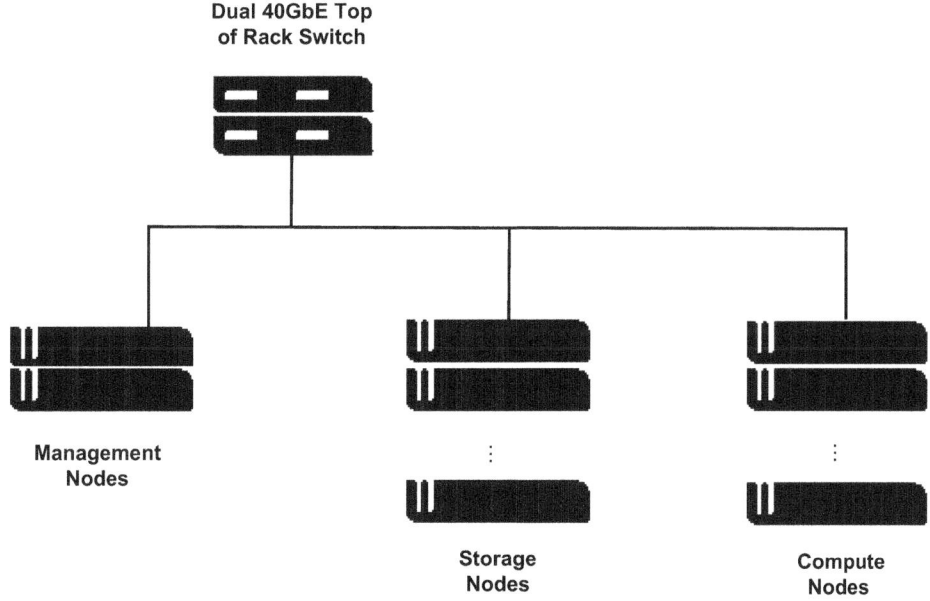

Figure 18-1 High-level cluster overview

CHAPTER 18
An Asymmetric Approach to Hadoop

As you can see in Figure 18-1, the solution consists of ProLiant Apollo 4200 servers and Apollo XL170r nodes connected to redundant HPE 5940 40 GbE top of rack switches. The Apollo 4200 servers are the data nodes, currently supporting 16–4 TB drives each, with the potential to grow to 24 Large Form Factor (LFF) drives. The Apollo 2000 chassis holds the 4 dense XL170r compute nodes, each with 28 cores and 256 GB RAM. These nodes support additional Solid State Disk (SSD) as needed for caching for some workloads. The DL360 Gen 9 servers act as the management nodes, hosting HPE's Cluster Management Utility, HPE Smart Update Manager, and Hadoop management tools. The following is more details on each server type:

- 3x Apollo 4200 servers with: (storage nodes)

    Dual E5-2640v4 processors

    128GB RAM

    Dual 40Gb QSFP+ ports

    16x 4TB drives

    Embedded P840ar/2GB controller

    Separate controller and drives for boot

    Operating System: Linux distribution compatible with the Hadoop distribution

    iLO Advanced licensing

    HPE Cluster Management Utility licensing

- 3x Apollo 2000 chassis, each containing 4 x XL170r nodes with the following specs: (compute nodes)

    Dual E5-2680v4 processors

    256GB RAM

    Dual 10Gb SFP+ ports (2 servers had 4 x 10G ports for external data access)

    2x 480GB SSDs (4 nodes with 4x 480GB SSDs for cache layer)

    B140i onboard SATA RAID controller

    Operating System: Linux distribution compatible with the Hadoop distribution iLO Advanced licensing

    HPE Cluster Management Utility licensing

- 2x DL360 Gen 9 servers each with: (management nodes)

    Dual E5-2640v4 processors

    128GB RAM

Dual 10Gb SFP+ ports

8 x 900GB SAS 10K drives

P440ar controller

Operating System: Linux distribution compatible with the Hadoop distribution

iLO Advanced licensing

HPE Cluster Management Utility licensing

- Dual HPE 5940 40GbE switches
- A single HPE 5900AF 1GbE switch for management
- Operating system and Hadoop licensing were purchased directly from the independent software vendors

Figure 18-2 depicts the front and back of the rack diagram for this cluster.

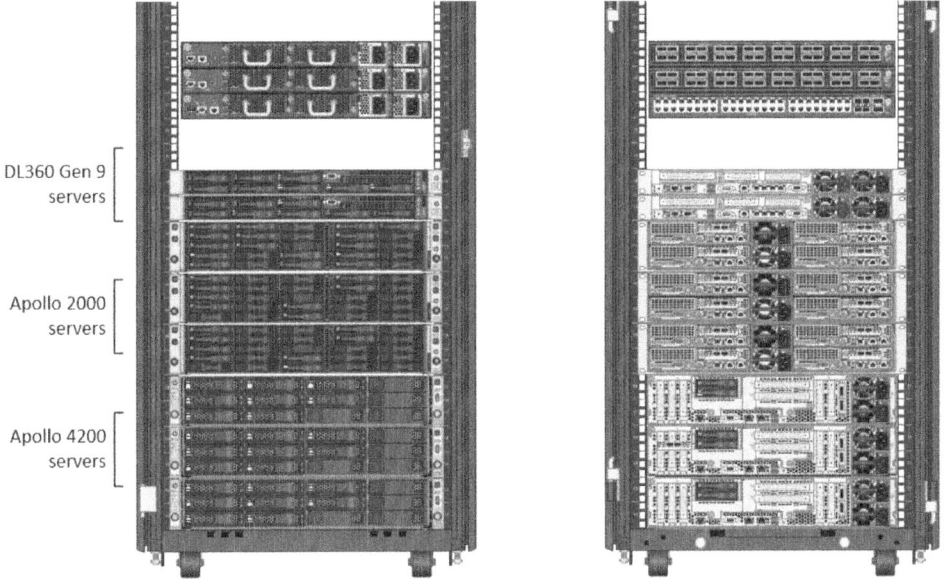

Figure 18-2 Starting Point of Cluster, consuming 17U

How did we arrive at this solution?

This section covers some of the thinking involved in determining this architecture.

Compute platform selection

Based on the considerations discussed earlier, the asymmetric architecture using Apollo servers was selected for the following reasons:

1. The Apollo 4200 has the most storage density in a 2U rackmount form factor in the HPE portfolio, while supporting the E5-2600v4 line and potential growth in networking. This was a perfect fit for the storage requirements in this build.

2. For the compute node, the Apollo 2000 offers the greatest compute density, with 4 dual-socket compute nodes in just 2 rack units, allowing up to 176 cores and 2 TB of RAM in just 2U.

3. By separating the compute and storage, the asymmetric WDO solution was able to use Hadoop YARN container "labels" capabilities, which allowed the flexibility to dedicate Hadoop or non-Hadoop workloads to their own compute nodes. Taking advantage of this functionality, we added solid state drives (SSDs) to some compute nodes for classical ETL as well as TEMP/scratch work for some SQL-centric workloads.

4. The 40 GbE network switch is a critical component because it overcomes the physical separation of storage and compute nodes by ensuring that the network would not be a bottleneck. Each of the 3 Apollo 4200 data nodes features an 80 Gb/s link into the HPN 5940 switches. Each of the 12 Compute nodes offered a 20 Gb/s link into the same HPN5940 switches providing a wealth of dedicated network bandwidth to the Hadoop cluster. All cluster nodes utilized Linux channel-bonding as well as Jumbo Frames (9K) for all IO traffic.

5. Using all X86 ProLiant servers for data nodes, compute nodes, and management nodes simplifies maintenance and management across the environment. This limits the number of teams that need to continually work together on varying server and storage environments.

There are additional ways these needs could have been met:

1. Traditional Hadoop architecture using DL380 Gen 9—This is the traditional Hadoop solution using a single x86 node to host both the storage and compute required. Although this option was considered, its reliance on symmetrically growing both compute and storage can be a limitation for some use cases. This model can leave stranded resources when compute is added each time storage needs to grow and vice versa.

2. Apollo 4500 server in place of the Apollo 4200—The Apollo 4510, 4520, and 4530 servers are also storage dense options for the data nodes. However, the Apollo 4510 offers even greater storage density but at the cost of the compute-storage ratio per node. This server is recommended in cases where Hadoop can utilize tiering, using the Apollo 4510 as the archive tier. The Apollo 4520 and Apollo 4530 do not offer equal storage density to the Apollo 4200, thus making this the perfect compute-storage balance for this use case.

3. SAN storage—The current incumbent environment featured a scale-out RDBMS that was deployed using SAN storage instead of recommended DAS storage.

4. Traditional Server and SAN environment—Neteeza is usually deployed with its own dedicated storage.

Working through these alternatives allowed us to identify the ideal solution for this specific application.

Compute node design

The key workloads at play were critical in designing the compute nodes in this architecture. Typically, the XL170r requires only boot drives because the primary data storage and any caching will be performed on the Hadoop Distributed File System (HDFS) or storage node layer.

In this case, four of the compute nodes required SSDs for integrating with legacy ETL tools as well as providing a TEMP area used for SQL join processing. Traditional Hadoop architectures would not allow for this change to only a few of the total nodes; however, using YARN container labels in this asymmetric architecture created this flexibility. It was simple to add two additional 960 GB SSD for a cache layer to four of the compute nodes. This made for a very cost-effective means of supporting this single workload, rather than adding unnecessary SSDs or compute nodes.

Drive selection

Careful consideration went into selecting the drives for the Apollo 4200 storage nodes. While the requirement was 40 TB per node, we ensured that there were sufficient disks to guarantee sufficient performance. This led to the selection of 16 x 4TB 12G SAS HDD, for 64 TB of storage per node, which is sufficient for growth. There are still eight slots left for additional storage per nodes before adding nodes.

This cluster will also use Apache Drill (apache.org/drill). This may require the addition of SSD for caching at the storage node layer, in order to ensure performance meets all SLAs. Drive slots are open for this purpose in the future.

Network setup

Performance was a major focus in developing this solution. Because the drives have been removed from the compute nodes, the network connecting the compute nodes to storage nodes is of even greater importance. This led to the selection of 40 GbE networks.

Each data storage node has dual 40 GbE connections to the top of rack switch connected via 40GbE Quad Small Form-factor Pluggable (QSFP+) copper Direct Attach Copper (DAC) cables. Jumbo Frames of 9K MTUs are used so that significantly less packet fragmenting occurs. Jumbo Frames means that 6x less system interrupts would occur on each Linux node's network device driver for increased scaling as compared to the use of 1500 byte MTUs. This allows data to travel without the network causing a bottleneck.

Since the compute nodes do not have the same high network requirements, these nodes each have dual 10 GbE ports, connecting to the top of rack switch via 40 GbE splitter cable. This cable has 40 GbE connections at the switch, breaking out into 4x 10 GbE connections to each node. This saves port count at the top of rack switch because each Apollo 2000 requires only 2x 40 GbE ports for all of its 4x XL170r nodes. This will be shown in an upcoming figure.

Management tool selection

The universal use of ProLiant servers with iLO4 ensured that management and monitoring of this environment remain simple. Additionally, HPE Cluster Management Utility licensing is included. This tool provides detailed performance and uses graphic diagrams, giving immediate insight into the health and status of each cluster. It also maintains a repository of golden images used for cloning to sets of nodes in the cluster, as well as scalable provisioning. HPE CMU has both a Graphical User Interface (GUI) and Command Line Interface (CLI) to ensure simplified use for different management styles. This licensing was included to ensure that the management of the cluster is simple, both from the start of provisioning and throughout the lifecycle of the cluster.

Management node design

The Linux distribution used does not require an independent Name Node because the metadata is spread across the cluster, typically on the dedicated storage nodes (Apollo 4200 servers). This provides built-in high availability for the traditional Name Node and reduces the number of dedicated management nodes required.

In this build, only two management nodes were required. One will host HPE's management tools: HPE Cluster Management Utility (CMU) and HPE Smart Update Manager (SUM). The second node will host the Hadoop GUI management tool and related functionality.

Proactive care advanced selection

Lastly, 3-year 24x7 Proactive Care Advanced support packs with Defective Media Retention were included for each component of the cluster. This ensures HPE's highest support level, including proactive credits for support throughout updates and an Account Support Manager (ASM). HPE support also works with third-party software manufactures to solve any issues that could be operating system related. These are critical factors in this mission critical cluster because as it allows you to work through a partnership with HPE to ensure uptime.

Figure 18-3 depicts the components of the cluster described earlier.

Ideal Platforms for Optimizing IT Workloads

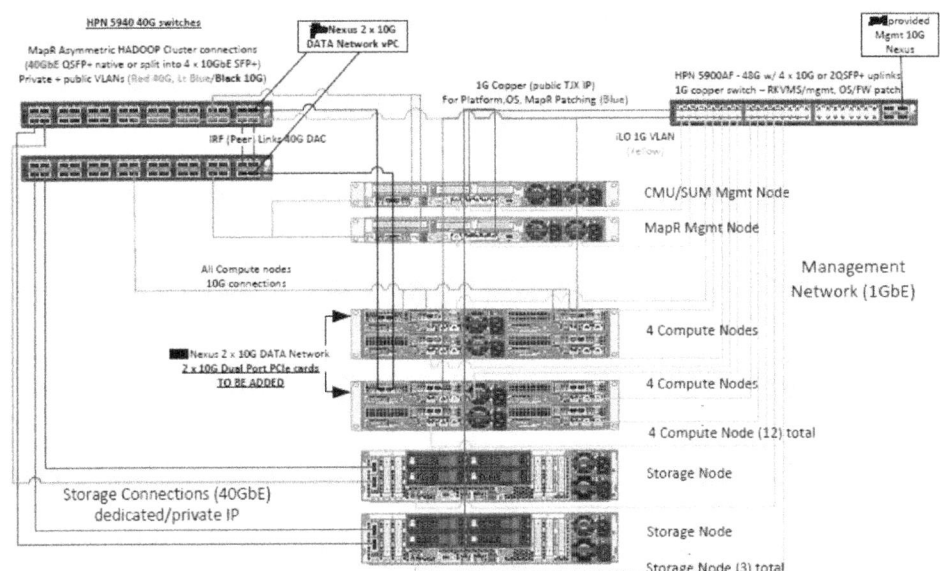

Figure 18-3 The components of the cluster

The next section covers the list of components for this cluster.

Hardware inventory

Table 18-1 shows the Bill of Material (BOM) for a single Hadoop cluster.

Table 18-1 BOM for Asymmetric Hadoop Architecture

		Factory Express Level 4 Services
1	ZU715A	HPE Virtual Rack
1	HA454A1-000	HPE FE Solution Package 4 SVC
3	HF482A1	HPE Factory Express Complex Custom SVC
		Management Nodes
0		HP DL360 Gen9 8SFF CTO Server [#4]
2	755258-B21	HP DL360 Gen9 8SFF CTO Server
2	755258-B21 ABA	US—English localization
2	818176-L21	HPE DL360 Gen9 E5-2640v4 FIO Kit
2	818176-B21	HPE DL360 Gen9 E5-2640v4 Kit
16	836220-B21	HPE 16GB 2Rx4 PC4-2400T-R Kit
16	785069-B21	HP 900GB 12G SAS 10K 2.5in SC ENT HDD
2	749974-B21	HP Smart Array P440ar/2G FIO Controller

CHAPTER 18
An Asymmetric Approach to Hadoop

Table 18-1 Continued.

		Factory Express Level 4 Services
2	779799-B21	HPE Ethernet 10Gb 2P 546FLR-SFP+ Adptr
2	734807-B21	HP 1U SFF Easy Install Rail Kit
2	764646-B21	HPE DL360 Gen9 Gen10 Serial Cable Kit
4	720478-B21	HPE 500W FS Plat Ht Plg Pwr Supply Kit
2	764636-B21	HP DL360 Gen9 SFF Sys Insght Dsply Kit
2	389692-B21	HP Customer Defined RAID Setting SVC
2	BD505A	HPE iLO Adv incl 3yr TSU 1-Svr Lic
2	469776-715	HP Add Generic Zpkg Kit
2	HA454A1-001	HPE FE Proliant Svr Pkg 4 SVC
1	HA114A1	HP Installation and Startup Service
2	HA114A1 5A0	HPE Startup Entry 300 Series OS SVC
1	H8B36A3	HPE 3Y Proactive Care Adv 24x7 wDMR SVC
2	H8B36A3 R2M	HPE iLO Advanced Non Blade—3 yr Support
2	H8B36A3 TT5	HPE ProLiant DL360 Gen9 Support
1	BD477A	HPE Insight CMU Media
		Storage Nodes
0		HP Apollo 4200 Gen9 24LFF CTO Svr [#3]
3	808027-B21	HP Apollo 4200 Gen9 24LFF CTO Svr
3	808027-B21 ABA	US—English localization
3	830728-L21	HPE Apollo 4200 Gen9 E5-2640v4 FIO Kit
3	830728-B21	HPE Apollo 4200 Gen9 E5-2640v4 Kit
24	836220-B21	HPE 16GB 2Rx4 PC4-2400T-R Kit
3	806564-B21	HP Apollo 4200 Gen9 2SFF and 2FHHL Kit
6	804575-B21	HP 80GB 6G SATA RI-2 SFF SC SSD
48	833928-B21	HPE 4TB 12G SAS 7.2K 3.5in MDL LP HDD
3	838827-B21	HPE SAS H240 FIO Ctlr Mode for Rear Strg
3	838823-B21	HPE Apollo 4200 Gen9 H240 Rear Cable Kit
3	726907-B21	HP H240 Smart HBA
3	764285-B21	HP IB FDR/EN 40Gb 2P 544+FLR-QSFP Adptr
6	720620-B21	HPE 1400W FS Plat Pl Ht Plg PS Kit
3	806562-B21	HP Apollo 4200 Gen9 Redundant Fan Kit
3	806565-B21	HP Apollo 4200 Gen9 iLO Mgmt Prt Kit
3	822731-B21	HP 2U Shelf-Mount Adjustable Rail Kit

Table 18-1 Continued.

		Factory Express Level 4 Services
3	822640-B21	HP Apollo 4200 Gen9 FIO Strap Ship Brkt
3	389692-B21	HP Customer Defined RAID Setting SVC
3	E6U64ABE	HPE iLO Adv incl 3yr TSU E-LTU
3	HA454A1-001	HPE FE Proliant Svr Pkg 4 SVC
3	469776-715	HP Add Generic Zpkg Kit
1	H8B36A3	HPE 3Y Proactive Care Adv 24x7 wDMR SVC
3	H8B36A3 R2M	HPE iLO Advanced Non Blade—3 yr Support
3	H8B36A3 YSN	HPE Apollo 4200 Support
1	HA114A1	HP Installation and Startup Service
3	HA114A1 58Y	HPE Startup Apollo 2000/4200 SVC
		Compute Nodes
0		HP Apollo r2600 24SFF CTO Chassis [#3]
2	798153-B21	HP Apollo r2600 24SFF CTO Chassis
2	798153-B21 ABA	US—English localization
4	800059-B21	HP Apollo 2000 FAN-module Kit
8	798155-B21	HP ProLiant XL170r Gen9 CTO Svr
8	850314-L21	HPE XL1x0r Gen9 E5-2680v4 FIO Kit
8	850314-B21	HPE XL1x0r Gen9 E5-2680v4 Kit
64	805351-B21	HPE 32GB 2Rx4 PC4-2400T-R Kit
8	726907-B21	HP H240 Smart HBA
8	779799-B21	HPE Ethernet 10Gb 2P 546FLR-SFP+ Adptr
16	832414-B21	HP 480GB 6G SATA MU-2 SFF SC SSD
8	798178-B21	HP XL170r/190r LP PCIex16 L Riser Kit
8	798180-B21	HP XL170r FLOM x8 R Riser Kit
8	798192-B21	HP XL170r/190r Dedicated NIC IM Board Ki
8	798207-B21	HP XL170r Gen9 Mini-SAS H240 Cbl Kit
8	389692-B21	HP Customer Defined RAID Setting SVC
8	E6U64ABE	HPE iLO Adv incl 3yr TSU E-LTU
4	720620-B21	HPE 1400W FS Plat Pl Ht Plg PS Kit
2	740713-B21	HP t2500 Strap Shipping Bracket
2	822731-B21	HP 2U Shelf-Mount Adjustable Rail Kit
2	469776-715	HP Add Generic Zpkg Kit
2	HA454A1-003	HPE FE Blade Svr Pkg 4 SVC

CHAPTER 18
An Asymmetric Approach to Hadoop

Table 18-1 Continued.

			Factory Express Level 4 Services
1	H8B36A3		HPE 3Y Proactive Care Adv 24x7 wDMR SVC
8	H8B36A3	R2M	HPE iLO Advanced Non Blade—3yr Support
2	H8B36A3	YHE	HPE Apollo 2000 Support
1	HA114A1		HP Installation and Startup Service
2	HA114A1	58Y	HPE Startup Apollo 2000/4200 SVC
			Compute Nodes (with additional cache layer)
0			HP Apollo r2600 24SFF CTO Chassis [#1]
1	798153-B21		HP Apollo r2600 24SFF CTO Chassis
1	798153-B21	ABA	US—English localization
2	800059-B21		HP Apollo 2000 FAN-module Kit
4	798155-B21		HP ProLiant XL170r Gen9 CTO Svr
4	850314-L21		HPE XL1x0r Gen9 E5-2680v4 FIO Kit
4	850314-B21		HPE XL1x0r Gen9 E5-2680v4 Kit
32	805351-B21		HPE 32GB 2Rx4 PC4-2400T-R Kit
4	726907-B21		HP H240 Smart HBA
4	779799-B21		HPE Ethernet 10Gb 2P 546FLR-SFP+ Adptr
16	832414-B21		HP 480GB 6G SATA MU-2 SFF SC SSD
4	798178-B21		HP XL170r/190r LP PClex16 L Riser Kit
4	798180-B21		HP XL170r FLOM x8 R Riser Kit
4	798192-B21		HP XL170r/190r Dedicated NIC IM Board Ki
4	798207-B21		HP XL170r Gen9 Mini-SAS H240 Cbl Kit
4	389692-B21		HP Customer Defined RAID Setting SVC
4	E6U64ABE		HPE iLO Adv incl 3yr TSU E-LTU
2	720620-B21		HPE 1400W FS Plat Pl Ht Plg PS Kit
1	740713-B21		HP t2500 Strap Shipping Bracket
1	822731-B21		HP 2U Shelf-Mount Adjustable Rail Kit
1	HA454A1-003		HPE FE Blade Svr Pkg 4 SVC
1	469776-715		HP Add Generic Zpkg Kit
1	H8B36A3		HPE 3Y Proactive Care Adv 24x7 wDMR SVC
4	H8B36A3	R2M	HPE iLO Advanced Non Blade—3 yr Support
1	H8B36A3	YHE	HPE Apollo 2000 Support
1	HA114A1		HP Installation and Startup Service
1	HA114A1	58Y	HPE Startup Apollo 2000/4200 SVC

Table 18-1 Continued.

		Factory Express Level 4 Services
		40GbE Data Switch and Cabling
2	JH396A	HPE FF 5940 32QSFP+ Switch
2	JG326A	HPE X240 40G QSFP+ QSFP+ 1m DAC Cable
6	JG328A	HPE X240 40G QSFP+ QSFP+ 5m DAC Cable
8	JG331A	HPE X240 QSFP+ 4x10G SFP+ 5m DAC Cable
4	JC680A	HPE 58x0AF 650W AC Power Supply
4	JC680A B2B	JmpCbl-NA/JP/TW
4	JG552A	HPE X711 Frt(prt) Bck(pwr) HV Fan Tray
2	HA454A1-021	HPE FE Strg and Ntwking Pkg 4 SVC
2	469776-715	HP Add Generic Zpkg Kit
2	H8B36A3 ZXP	HPE 5940 Fixed 48G Support
		1 GbE Management Switch
1	JG510A	HPE 5900AF 48G 4XG 2QSFP+ Switch
2	JC680A	HPE 58x0AF 650W AC Power Supply
2	JC680A B2B	JmpCbl-NA/JP/TW
2	JC682A	HPE 58x0AF Bck(pwr) Frt(prt) Fan Tray
1	HA454A1-021	HPE FE Strg and Ntwking Pkg 4 SVC
1	469776-715	HP Add Generic Zpkg Kit
1	HA114A1	HP Installation and Startup Service
1	HA114A1 5RN	HPE Top of Rack Startup SVC
1	H8B36A3 R4W	HPE 5900AF-48 2QSFP Switch Support
		Cabling for 1 GbE Management Connections
36	C7536A	HP Ethernet 14ft CAT5e RJ45 M/M Cable
		Management Licensing and Support
16	BD476A	HPE Insight CMU 3yr 24x7 Flex Lic
1	H8B36A3	HPE 3Y Proactive Care Adv 24x7 wDMR SVC
16	H8B36A3 4QC	HPE Insight CMU 3yr 24x7 FlexLic SW Supp
		Proactive Care Advanced Support Credits
110	U4VT2AS	HPE PCA Proactive Credits Per Year SVC

The next section covers the implementation of this mission critical cluster.

CHAPTER 18
An Asymmetric Approach to Hadoop

Implementation project plan

The final critical element of this environment was simplifying the deployment by incorporating Factory Express Level 4 services at HPE's Americas Integration Center (AIC) for extensive factory services.

In this case, servers that were integrated were installed on-site in existing racks. If existing racks were not used, the AIC would rack, cable, and setup the infrastructure in their build center in order to perform the OS installation and networking setup. They can then unrack and send the servers separately before it is delivered. This is beneficial for scenarios in which existing data center infrastructure must be used.

These services also include the installation of the Linux operating system in HPE's AIC. By providing the specific operating system image with any required settings, HPE can provide the servers fully built with standards in place.

Lastly, the service SKUs listed also include the network setup which includes, IP addresses, VLANs, and so on. Having these elements allows simple incorporation into the existing environment eliminating manual setup.

Once the factory services are complete and the solution is delivered, HPE installation services and complex on-site services allow for the rack and stacking, setup, initial connectivity testing, and setup of HPE CMU. This also includes collaboration with the Hadoop team for the installation and setup of Hadoop.

Through the combination of these services, a fully packaged solution is delivered, with hardware and software installed, and configuration complete, by the time they begin moving workloads onto the new infrastructure.

Summary

This design was intended to meet current storage and compute requirements and grow as needed for this Hadoop cluster.

Some of the key aspects of this solution related to future growth were as follows:

- HPE CMU was included for simplified management for the current infrastructure, with easy understanding of the performance and utilization factors. This also allows for simplified adding of more nodes to the same cluster.
- Each server, both compute and storage nodes, maintains space for storage growth allowing growth within each node before adding to the total nodes in the cluster.
- The network switches were designed with room for future growth as well, allowing these clusters to grow within the network domain.

- The initial deployment was intended for use in two scenarios: the test cluster and the production cluster. Due to its building-block and intrinsically flexible design, the same basic cluster design can be easily repeated for other uses by simply adding and removing storage and compute nodes based on the specific cluster needs.

The end result is a cluster that was designed to meet specific needs as well as support growth and was delivered in a way that it was up-and-running quickly.

19 Hybrid IT: The "Right Mix"

INTRODUCTION

This chapter covers a solution that includes the hardware infrastructure, software tools, professional services, and educational resources required for the creation of, and migration to, a hybrid IT environment.

Hybrid IT is an information technology strategy that uses a mix of public, private, and managed infrastructure as a single collection of resources. These resources may represent very different types of infrastructure including traditional enterprise IT architectures, traditional virtualization, containers, private cloud, public cloud, Software as a Service (SaaS), managed solutions that are externally hosted, managed solutions that are internally hosted.

Hybrid IT enables an organization to provision the right infrastructure for the right workload. Workload placement decisions can be manual, semi-automated, or fully automated. Just a few of the workload placement criteria an organization might use are price considerations, performance requirements, security concerns, governance rules, and service levels. The result is an organization that begins to achieve the "Right Mix" of IT modalities in order to satisfy its goals.

Benefits of hybrid IT adoption include cost-effectiveness, reduced risk, and increased flexibility. These are achieved through careful selection of resource providers to the hybrid IT environment and efficient workload placement.

Some requirements of a well-designed hybrid IT environment include:

- Scalable Hardware Architecture—By choosing an easily scalable architecture, IT can quickly respond to the demands of internal customers and external pressures, increasing capacity while reducing downtime.
- Open Service Catalog and Orchestration—A truly open service catalog and orchestration solution is one that includes integrations for third-party products and services, can be easily modified, and can be extended to include new integrations. Integrations with third-party products and services should be publicly available and customizable by the end user. The service catalog and orchestration provider may still maintain proprietary rights over them, but should not require input or permission from the other providers or organizations for integration.

- Application Programming Interface (API), Accessible Hardware and Software—Well-documented APIs will provide integration with existing systems and expansion through new products and services. APIs should be present at the hardware and software level, and allow access to low and high-level functionality (installation-to-configuration-to-implementation).

Solution overview

This solution was designed to help this user achieve the "Right Mix" of IT resources while enabling more flexibility, more visibility, and more freedom in a relatively short period of time. This was achieved by using the latest release of the HPE Helion CloudSystem in conjunction with HPE's best-in-class cloud hardware infrastructure, application transformation services for new and existing apps, and education for operators and end users. Organizations using this solution will allow the end user the flexibility to use multiple public and private cloud providers as well as consume traditional architectures as infrastructure as a service (IaaS).

The benefits of this solution include:

- **Faster, easier, and more flexible IT infrastructure management**—The solution natively supports the ability to provision and manage resources with many public and private cloud providers, as well as containers and traditional virtualized/bare-metal architectures out-of-the-box. This enables new flexibility in design, deployment, and operations across technology modalities. The end user needs only to focus on processes and workloads, and the solution will automate the provisioning process to deploy workloads where they fit best.

- **Enhanced networking, storage, servers, lifecycle management, and security features**

  —Operations staff are often overloaded with manual processes, unplanned requests, demands for new features, and increased requirements. This solution uses software-defined infrastructure and IaaS software solutions that extend existing IT service offerings while adding new IT services.

- **Powerful tools to build and deploy apps across multiple clouds**—Included in the solution are the tools that operations teams can use to provide developers with the ability to build, test, deploy, and manage cloud-native apps and services across private and public cloud environments, and between suppliers. By enabling the deployment of a single code base to multiple clouds both public and private, organizations will realize improved efficiencies and time to market for cloud-native applications and services.

- **Continually updated cloud marketplace and service designs**—The solution includes a customizable, intuitive, self-service portal with a "catalog shopping" experience. Catalog offerings to deploy hundreds of architectures from dozens of manufacturers are included and can be customized easily into thousands of combinations of infrastructures and applications. Deploying these advanced infrastructure offerings and complex application services is enabled by using the built-in graphical service designer.

Solution components

Several components are used together in varying configurations to achieve the "Right Mix" of resources to create a hybrid environment. These components include servers, storage, networking, traditional, virtualized, containers, private cloud, and public cloud. The Hybrid IT solution described here spans all these components, providing comprehensive provisioning, management, insight, and control. It uses HPE's Hyper Converged 380, CloudSystem Enterprise (only), Education services, and consulting services to achieve this efficiently and in a relatively short period of time. Figure 19-1 provides a high level overview of the solution components, including assembly and implementation.

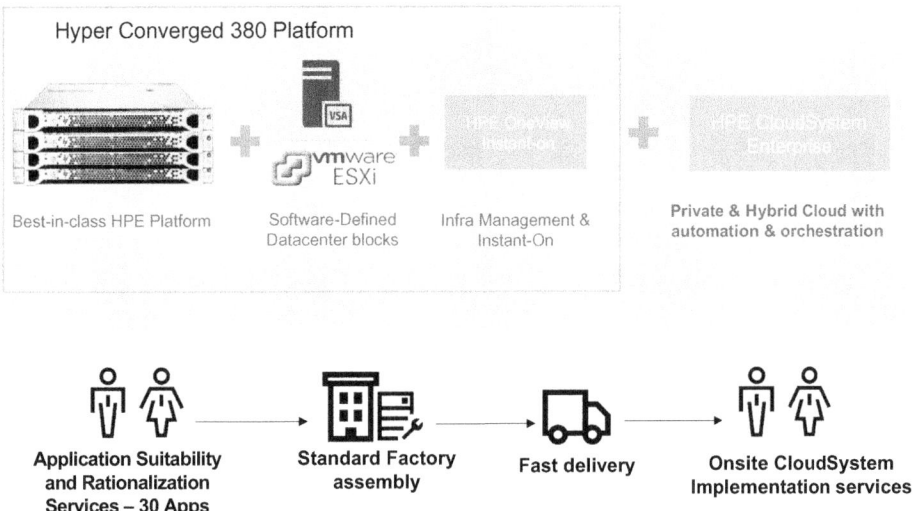

Figure 19-1 High-level "Right Mix" solution

Hyper Converged 380

The base of the solution is the HPE Hyper Converged 380 (HC380). The HC380 provides a base infrastructure solution with pre-integrated virtualized compute, network, and storage. The HPE Hyper Converged 380 combines HPE ProLiant DL380 servers, HPE StoreVirtual software-defined storage, VMware virtualization software, and HPE management tools to deliver a system that is quick to install, easy to manage, and easy to expand. The result is a "VM Vending Machine" that is as easy to manage as it is to consume. If needed, it can be expanded quickly from as little as 2 nodes to up to 16 nodes with a simple user interface. Management software automatically recognizes and integrates new nodes into the cluster. In addition, it will perform live, non-disruptive firmware management from the integrated user interface (Figure 19-2).

CHAPTER 19
Hybrid IT: The "Right Mix"

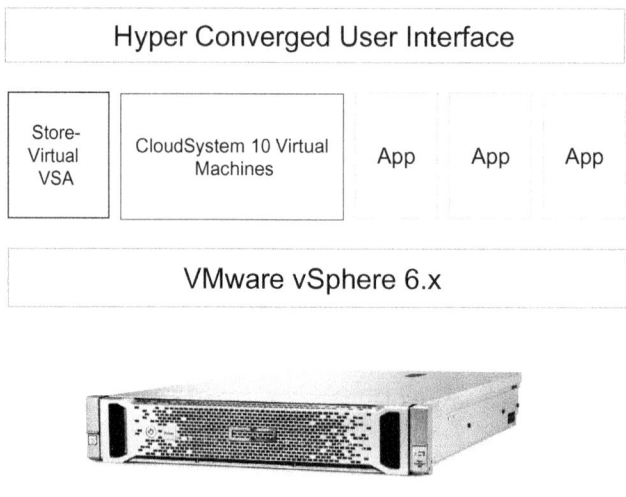

Figure 19-2 HPE Hyper Converged 380

The HC380 supports Intel's 14 nm and 22 nm Broadwell/Haswell processors and all flash software-defined storage from HPE. Storage-only blocks can be added to increase capacity in existing nodes in a cluster, and storage-only nodes can be used to scale storage independently from compute.

Helion CloudSystem 10 Enterprise

The software solution running on top of the hardware is the HPE CloudSystem 10 Enterprise. CloudSystem 10 is a software solution for private and hybrid cloud, delivering automation and orchestration of traditional and cloud native workloads. The CloudSystem software is delivered as two packages:

- HPE Helion CloudSystem Foundation—targets simple private cloud scenarios, delivering IaaS and Platform as a Service (PaaS) on HPE hardware.

- HPE Helion CloudSystem Enterprise—enables a broader range of hybrid cloud scenarios, delivering IaaS, PaaS, and multi-IaaS management on HPE hardware.

HPE CloudSystem 10 Enterprise is depicted in Figure 19-3.

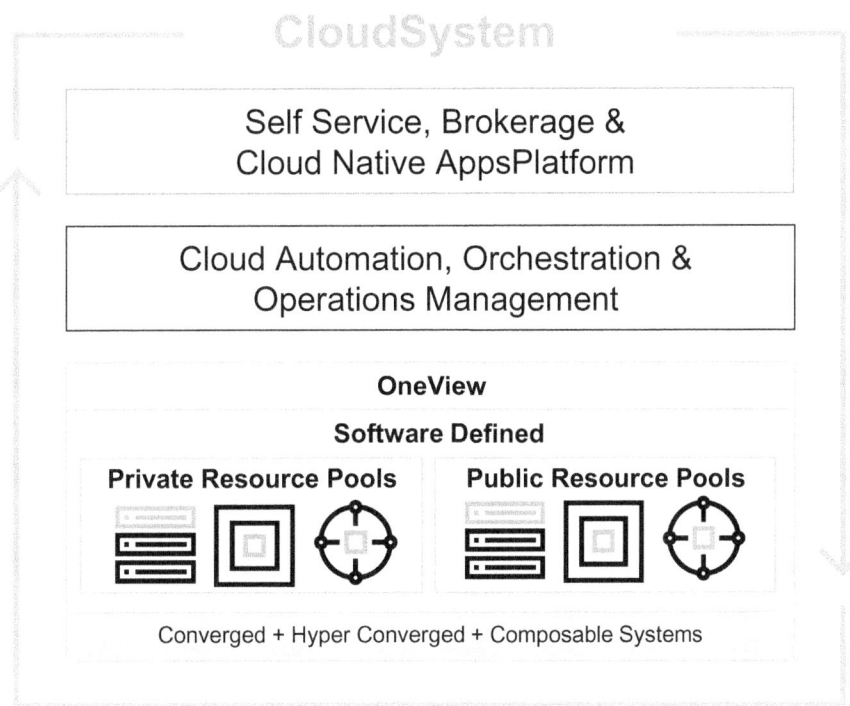

Figure 19-3 HPE CloudSystem 10 Enterprise components

The components of the Foundation edition include Helion OpenStack 4 and Helion Stackato 4. The components of the Enterprise edition include the Foundation edition plus Cloud Service Automation 4.6, and Operations Orchestration 10, all within a virtualized installation bundle. The Enterprise edition, which adds Cloud Service Automation service catalog and Operation Orchestration automation to Foundation, is also available without Foundation (Enterprise "only"). The Enterprise only variant is used in this solution.

The CloudSystem features a simplified licensing model for the components included as a single product that is licensed per server with unlimited virtualized instances per physical server plus five managed Open System Interconnection (OSI) for public cloud per license. The CloudSystem will run in a supported configuration on a broad range of hardware platforms in addition to the Hyper Converged 380 including the CS700, C7000, Synergy, and ProLiant.

Cloud Service Automation

The Cloud Service Automation (CSA) component of the CloudSystem is cloud management software that helps deploy hybrid services from multiple clouds or other sources. It is used to automate and simplify the deployment and management of several types of IT architectures in order to centralize hybrid IT services. End users can customize these services in an intuitive, self-service portal with a modern catalog shopping experience.

CHAPTER 19
Hybrid IT: The "Right Mix"

CSA helps eliminate vendor lock-in by leveraging the APIs or CLIs of each infrastructure vendor an IT operator includes in a service design. The service design can be made up of multiple hypervisors, multiple hardware vendors, private cloud software, and public clouds. CSA then offers its own API, CLI, and service catalog for interacting with the service designs. This enables the ability to deploy advanced infrastructure and complex multi-tiered applications very easily via web-based portal or programmatically. And since service designs are based on TOSCA, a vendor-neutral service design modeling standard, they are portable (Figure 19-4).

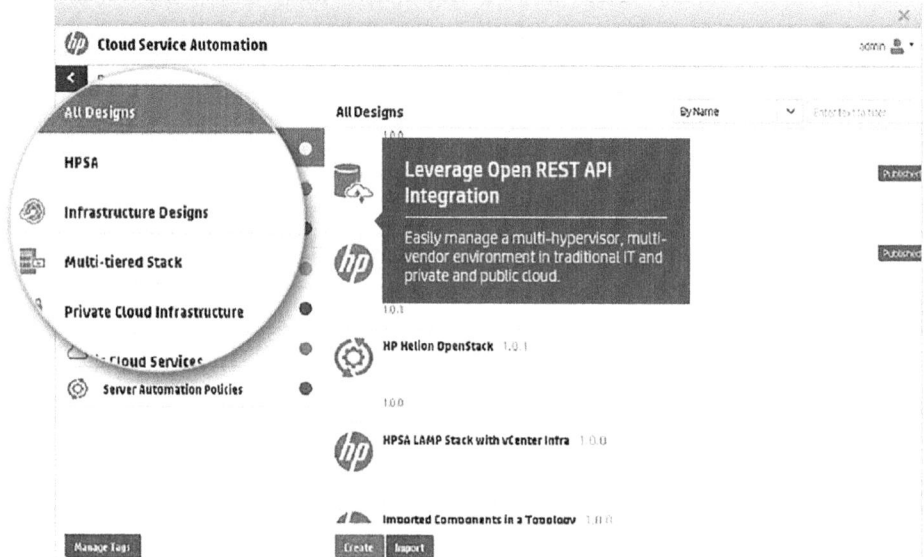

Figure 19-4 Cloud Service Automation

Operations Orchestration

The Operations Orchestration (OO) component of the CloudSystem is an IT process automation and run book software that improves service quality and lowers costs by eliminating latency between infrastructure and software teams, enforcing standards, delivering reports and audits, and more. OO will manage the installation, update, removal, and reporting of configurations, software, files, objects, and scripts. It is used to quickly and accurately execute processes and deploy infrastructure across multiple environments automatically, reducing errors in process execution. In addition, OO comes with 5000+ prebuilt operations and 80+ integrations so that users can automate multiple-vendor environments quickly and string together automated processes into orchestrated work flows. These workflows can be accessed via PowerShell, REST API, or CLI (Figure 19-5).

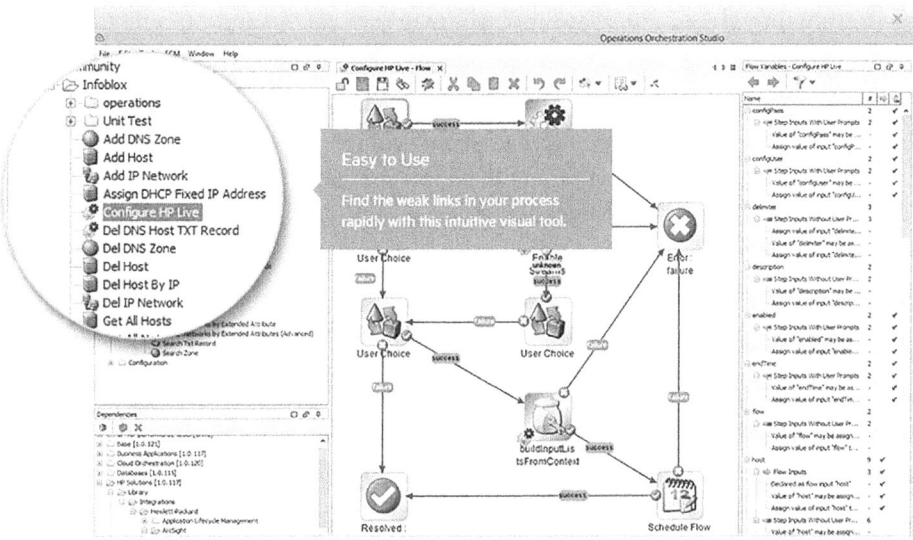

Figure 19-5 Operations Orchestration

Helion OpenStack

The Helion OpenStack component of the CloudSystem provides a highly configurable, enterprise-grade cloud IaaS platform based on open-source OpenStack technology with enterprise-level support. As HPE's enterprise-grade distribution of OpenStack software, Helion OpenStack is hardened and extensible to deliver leading open source cloud computing while adhering tightly to OpenStack software API standards and services (Figure 19-6).

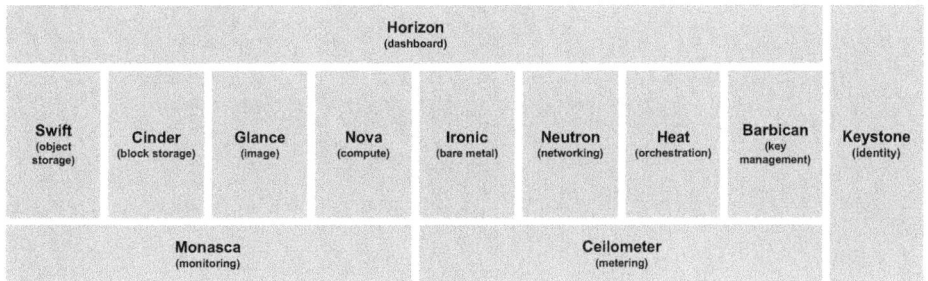

Figure 19-6 Helion OpenStack

Helion OpenStack includes the Helion Lifecycle Management product, providing an Ansible-based installation and configuration tool that gives users the ability to add or remove services independently via a web-based UI. In addition, users can provision new or replace defective infrastructure without impacting the running cloud, automate lifecycle management operation, and track and audit all changes made to the cloud.

The included Operations Console is a web-based GUI that offers the user access to data such as monitoring alarms data by service, performance data, compute nodes and instances lists, and a drilldown to Kibana for a deep look into logs. There is also a Business Logic Layer, a middleware component that serves as a single point of contact for the UI to communicate with OpenStack services such as Monasca, Nova, and others.

Helion Stackato

HPE Helion Stackato is an application platform with a focus on ease-of-code development while being simple for administrators to deploy in a multi-cloud environment. Helion Stackato can be deployed on private virtualized infrastructure, private cloud, or public clouds. Helion Stackato runs on various distributions of OpenStack as well as public clouds. The platform also supports a wide variety of programming languages, including native .NET support, Java, and newer languages such as Node.js, Python, Ruby, and more. Developers using Helion Stackato can work with source code or directly with Docker container images.

Figure 19-7 depicts this architecture.

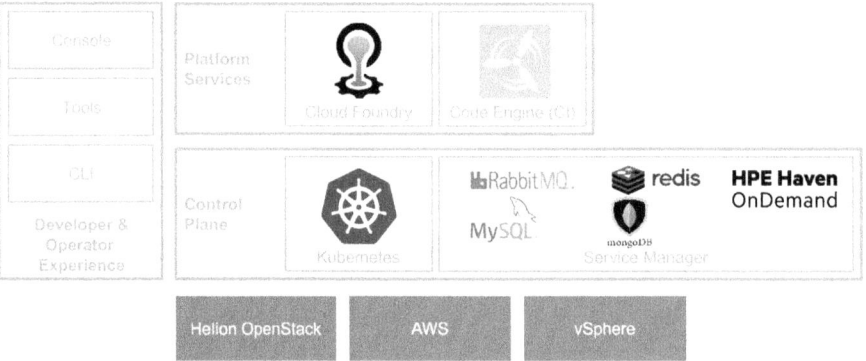

Figure 19-7 High-level Stackato Architecture

The use of Helion Stackato enables users to easily deploy cloud-native applications in a multi-cloud environment. This will accelerate the development of cloud-native applications by reducing the complexity of application deployment across development and operations. The use of Stackato in this solution also enables the user to easily scale cloud applications within a multitenant architecture using containers.

The next section describes the solution and includes the components of which the solution is composed.

 Note
Some of the software components covered in this chapter will be included in the spin-off and merger with Micro Focus. Information about this is found at the following URL: https://www.hpe.com/us/en/newsroom/news-archive/press-release/2016/09/1276021-hpe-accelerates-strategy-with-spin-off-and-merger-of-non-core-software-assets-with-micro-focus.html

Solution inventory

Digging into the details of this solution you will see that the hardware configuration of the compute nodes consists of six HPE HC380 utilizing the Intel® Xeon® Processor E5-2680 v4 (35M Cache, 2.40 GHz). The total number of cores offered as part of the solution is 168 cores, or 28 cores per host.

Memory offered via HPE DDR4 SmartMemory modules for a total of 3072 GB or 512 GB per host.

Storage is supplied by the HPE StoreVirtual software-defined SAN. Each node uses a combination of 480 GB SSDs and 900 GB 10K HDDs for a total of 26 TB usable storage as a single, software-defined iSCSI SAN with over 60,000 IOPS of performance.

Network connectivity is enabled via the HPE Ethernet 10 Gb 2-port 561T adapter. The adapter is a dual-port 10GBASE-T device featuring the Intel X540 10 Gb Ethernet solution in a PCIe 2.1 compliant form factor. The HPE 561T adapter delivers full line rate performance.

Figure 19-8 depicts the solution at a high level:

CHAPTER 19
Hybrid IT: The "Right Mix"

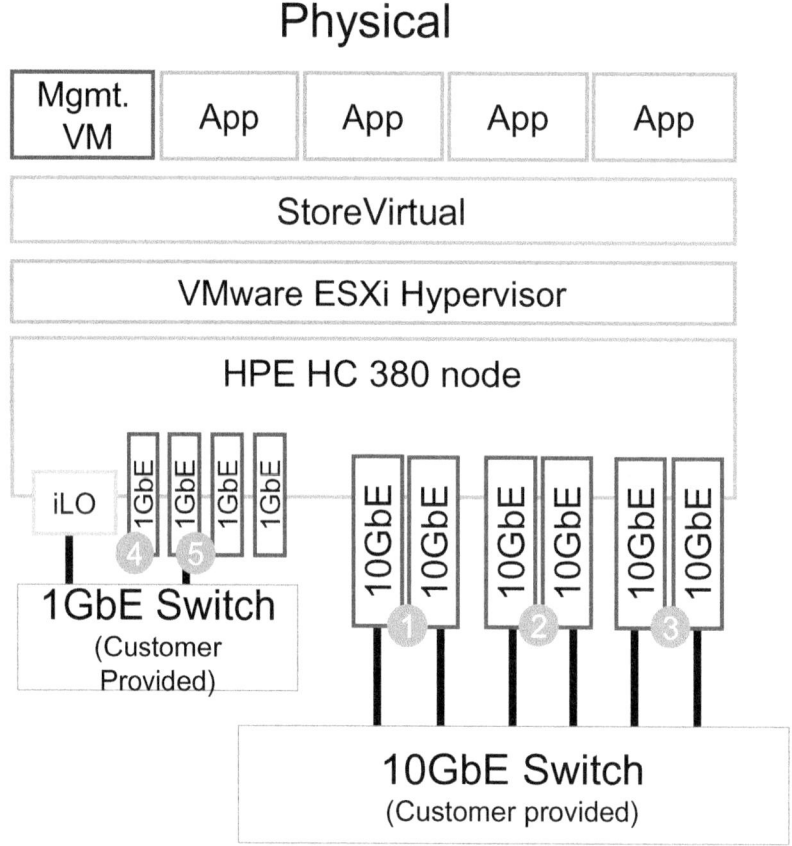

Figure 19-8 High-level solution diagram

Table 19-1 is the solution component listing, or Bill of Materials (BOM).

Table 19-1 Bill of Materials

Quantity	Part #		Description
6	P9D74A		HPE HC380 Cluster Node
6	P9D74A	003	HPE HC380 Cloud SW
6	817951-L21		HPE DL380 Gen9 E5-2680v4 FIO Kit
6	817951-B21		HPE DL380 Gen9 E5-2680v4 Kit
6	817951-B21	0D1	Factory integrated
96	805351-B21		HPE 32GB 2Rx4 PC4-2400T-R Kit
96	805351-B21	0D1	Factory integrated
72	785069-B21		HPE 900GB 12G SAS 10K 2.5in SC ENT HDD
72	785069-B21	0D1	Factory integrated
24	816985-B21		HPE 480GB 6G SATA MU-3 SFF SC SSD

Table 19-1 Continued.

Quantity	Part #		Description
24	816985-B21	0D1	Factory integrated
6	749974-B21		HPE Smart Array P440ar/2G FIO Controller
6	726897-B21		HPE Smart Array P840/4G Controller
6	726897-B21	0D1	Factory integrated
12	783009-B21		HPE DL380 Gen9 8SFF SAS Cable Kit
12	783009-B21	0D1	Factory integrated
6	786092-B21		HPE DL380 Gen9 8SFF H240 Cable Kit
6	786092-B21	0D1	Factory integrated
6	700699-B21		HPE Ethernet 10Gb 2P 561FLR-T Adptr
6	700699-B21	0D1	Factory integrated
12	716591-B21		HPE Ethernet 10Gb 2P 561T Adptr
12	716591-B21	0D1	Factory integrated
12	720479-B21		HPE 800W FS Plat Ht Plg Pwr Supply Kit
12	720479-B21	0D1	Factory integrated
6	768900-B21		HPE DL380 Gen9 Sys Insght Dsply Kit
6	768900-B21	0D1	Factory integrated
6	719073-B21		HPE DL380 Gen9 Secondary Riser
6	719073-B21	0D1	Factory integrated
6	733660-B21		HPE 2U SFF Easy Install Rail Kit
6	733660-B21	0D1	Factory integrated
6	666988-B21		HPE 2U Security Bezel Kit
6	666988-B21	0D1	Factory integrated
6	733664-B21		HPE 2U CMA for Easy Install Rail Kit
6	733664-B21	0D1	Factory integrated
6	758959-B22		HPE Legacy FIO Mode Setting
6	P9D85A		HPE HC380 Base SW Image 6.0 FIO Kit
1	H1K92A3		HPE 3Y Proactive Care 24x7 Service
6	H1K92A3	XW4	HPE HC380 Cluster Node Support
3	F9D69CAE		HPE CS10 Ent 1-Svr 1y S E-LTU
1	H1K92A3		HPE 3Y Proactive Care 24x7 Service
3	H1K92A3	TFL	HPE CS Enterprise 1-Server Support
1	HA124A1		HPE Technical Installation Startup SVC
1	HA124A1	5VB	HPE CloudSystem Enterpris SW Startup SVC
6	HA124A1	5Z0	HPE HyperConverged 380 Startup SVC
1	H8Q70A1 #501		HPE CloudSystem Accelerator Service

Education and application services that are part of the solution are detailed in Table 19-2 and later in this chapter.

Table 19-2 Education and application services

Service	Descriptions
H4C08S	HPE Helion CloudSystem Enterprise Training
H4S74S	HPE Cloud Services Automation Training
H4S75S	HPE Operations Orchestration Training
H0LP9S	HPE Hyper Converged Administration Training
H4C04s	HPE One View Administration Training
(SoW-based)	Application Workload Suitability Assessment for 30 Applications

Services and education

Both application transformation and education services were described as a critical part of the solution. HPE knows this because our customers validate market research on why cloud initiatives fail, some of which includes using the wrong technology, underutilization of cloud resources, and failure to change the operational model of the organization to enable cloud usage. HPE application transformation and education services can help reduce the risk associated with these issues.

HPE Application Transformation services—HPE Application Transformation services are geared toward helping customers with the various activities that support a medium-to-large-scale IT transformation to cloud. In this services engagement, HPE uses experience to enable successful "migration to cloud" projects. Using these services, you can choose any industry-standard private and public cloud platform and build a common transformation model.

HPE's phased approach toward cloud portability includes design, assessment, rationalization, and reporting. The following list summarizes the services:

- Future State Design—Work with stakeholders to consider and identify options of Hybrid IT deployment and configuration automation requirements.

- Assessment Finalization—Identify assessment criteria in correlation with established business priorities and goals.

- Application Assessment—Conduct a tool-based application assessment and data collection with workshops across business units in scope.

- Application Rationalization—Correlate data collected from business-unit-level workshops and use tool-based analysis and consulting to provide reports that will be used to create the final deliverable.

- Reporting and Wrap-up—Provide final deliverables that will include findings of application assessments across business units, transformation recommendations, and commercially recommended practices

The final deliverable includes an application assessment report with detailed findings. The contents of this deliverable will include:

- Dashboards of application assessment results
- End-state recommendations for applications
- Recommendation on next steps for application
- Application groups based on established criteria
- Suitability score for applications and application groups based on established operational mode

HPE Hybrid IT Education Services—Education for hybrid IT environment operators and administrators is key to the successful maintenance and growth of the solution. Included here is Cloud Operator, Cloud Architect, and Hyper Converged Administrator training.

Cloud Operator training enables you to use and customize the cloud software, as well as plan and prepare for future CloudSystem operations tasks. Administration tasks from creating and maintaining resources to executing real user scenarios are covered.

Cloud Architect training includes covering the orchestration engine to locate, run, monitor, and perform important administrative and operational tasks related to cloud provisioning and maintenance. You will learn how to interact with a variety of services including public cloud infrastructure providers, web services, e-mail providers, files systems, and so on.

Hyper Converged Administrator training is for infrastructure administrators and system engineers who are looking to learn Hyper Converged system installation, configuration, and administration techniques. This course provides information on installation, configuration, and management tasks for HPE Hyper Converged 380 system. The course covers pre-installation, installation, and configuration tasks for the HPE HC 380 system using HPE OneView InstantOn and administration of HPE HC 380 system with the Management User Interface to develop expertise in the steps which are necessary to install and configure the HPE HC 380.

Summary

There are many benefits realized by this Hybrid IT design including, but not limited to the following:

- Optimize the "Right Mix" of IT with a balance of existing and new internal and external services
- Empower a rising developer class to build apps and micro-services fast

CHAPTER 19
Hybrid IT: The "Right Mix"

- Deliver "always-on" services
- Auto-scale to maximize your business' growth potential
- Move from large, monolithic applications
- Reduce operational cost with automation
- Reduce Capital Expenditure with the appropriate resource utilization

This is an ideal solution for those who want to enable hybrid IT service delivery and get-up-and-running with cloud services quickly and efficiently. The solution's small footprint and the relatively lower price point of HPE Helion CloudSystem Foundation software make it useful as a launch pad for customers just getting started with hybrid cloud.

20 The Power of Connection: The Internet of Things

INTRODUCTION

Many technological advancements in computing over the years have focused on making existing capabilities more functional or effective, less expensive, or easier to use. However, other advancements, such as the advent of the Internet, have been more transformative, creating new areas of opportunity that did not exist before. With the explosion of connected devices, exponential growth in data, pervasive connectivity and mobility, among others, the technologies supporting the Internet-of-Things (IoT) stand together to be equally as transformative in businesses and society, unlocking new chains of value.

IoT includes the ability to provide some form of connectivity for any object, such as a machine, gather data collected from these objects, and then perform analysis at various levels to gain insight. Analysis can be performed near the object to produce analysis that can be acted upon in real time, or deep analytics performed to provide insight in the form of historical or predictive analytics. The capability to obtain and analyze data in various ways will transform and impact virtually every aspect of personal life, society, consumer, and industrial markets.

In this context, the world of IoT can be viewed from two main perspectives: Industrial (IIoT,) which has to do with business and operational considerations in areas such as retail, manufacturing, transportation, and energy, and Consumer (CIoT), which addresses personal devices and activities such as those in a connected home.

When using the term "IoT," this chapter does so from an industrial perspective in discussing the following:

- The Opportunity With Connected, Intelligent Things
- Common IoT Architecture Elements
- The Potential of a Connected Future

The first section highlights the potential that exists for IoT from a business perspective. Businesses will find that IoT can not only help improve current operations but also potentially open new business opportunities. This will help the reader's understanding of the purpose and value of IoT.

Next, this chapter addresses a high-level component and technological approach that can be used to define the high-level "common core" essential elements of an IoT architecture. This will help understand how IoT applications are built.

Lastly, a potential view of the future possibilities of the IoT world is considered.

The opportunity with connected, intelligent things

The term "connected device" has undergone significant change over the years. At the earlier advent of commercial computing devices, such as tape devices, terminals, printers, these could only be connected to a host computer using a limited number and type of proprietary or other "standard" physical and protocol connection methods such as Systems Network Architecture (SNA from IBM) and RS-232 serial communications.

As the computing industry has transformed, these types of connections have multiplied. While IoT has much to do with connecting an object to the Internet, any object can now be connected to have data retrieved from it and communicate with other objects. These connections can be localized, as in connecting a series of sensors on a machine, to network connections for local and remote connectivity.

In addition, the types of objects being connected have changed significantly. Machines or other objects that had no connectivity capabilities previously can now be fitted with sensors that can be connected as described earlier. Devices that would not have been considered for connection such as automobiles, clothing outfitted with sensors, location objects, hand-held devices, trains, treadmills, and refrigerators are all increasingly expected to have such connectivity.

Beyond the types of connections and devices, the intelligence of devices is changing. This means that while many connected objects will simply be connected so they can be monitored and/or controlled operationally, other devices will become more capable of performing analysis themselves. These devices may also be leveraging localized compute resources. In either case, it will become less necessary to rely on connectivity, compute, and data processing capability to be provided elsewhere such as within the traditional IT shop. This is known as edge computing.

As an example, consider the evolution of the automobile. The cars of today are essentially moving data centers with advanced connectivity and computational capabilities to not only monitor vehicle operations but also increasingly make analytical and autonomous decisions inside and outside the vehicle.

This world of connected, intelligent objects, whether machine or animate, and edge computing presents the opportunity to collect data from those objects, to have the objects themselves participate in, inform or decide actions, and to interact with each other for mutual beneficial operations and business insight.

While IoT includes such technological shifts, it also involves a shift in business thinking. With the potential for connectivity, access to further data sources in operations and within traditional IT, and analysis capabilities, changes to a company's business and operational approach and processes need to be considered. This can be as challenging as the technological considerations but also profitable in helping to realize new business value potential and competitiveness.

As companies begin their journey into IoT, many are looking first to operational benefit. Connecting existing or new factory-floor devices, hospital room equipment, transportation vehicles, or any other business device can allow more data collection for real-time or historical analysis. This can help to improve operational efficiencies and lead to more timely decisions.

As they progress in IoT experience, companies will find that the ability to connect intelligent devices, gather data, and move analysis closer to the objects themselves, or within the devices will produce the potential for new business opportunity, services, and customer value.

As an example consider how a retail store could realize additional competitive advantage and provide more customer value by using connected inventory in their stores, provide location services that promote products and target customers real time as they move through a store, link real-time store data to local and broader inventory, buying patterns, or many other in-house and external info sources to adjust to customer demands. Likewise, in every industry, similar business analysis as well as operational benefit can be considered.

This connectivity, access to data, and moving computational capabilities closer to the devices themselves is considered the Operational Technologies (OT) component of IoT. Coupling this data with traditional data center or IT elements provides additional opportunity for the operational and business benefits discussed. Realizing this value perspective for IoT provides a foundation for next considering the major components necessary to implement IoT applications.

Figure 20-1 depicts, at a high level, some of the opportunity that exists with IoT.

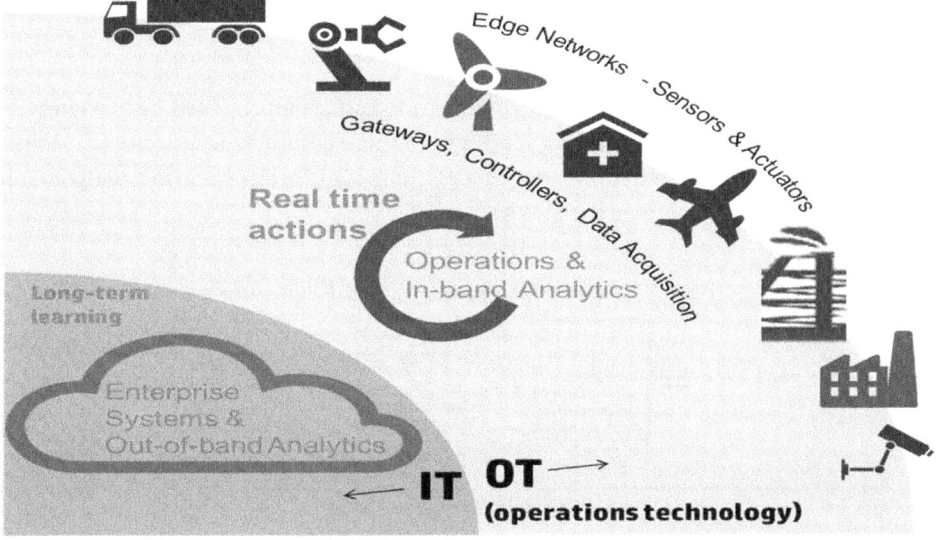

Figure 20-1 The opportunity for IoT—data and edge analytics

Common IoT architecture elements

To some degree, connecting various devices and gathering data is not new. Manufacturing companies have long used sensors, actuators and other connection technologies to monitor and control operational elements. An example would be gathering data to monitor and control a generator in a power plant.

What makes IoT unique and different from past connectivity efforts is the scope of device types, the many methods of connecting those devices, the locality of the devices and data, the volume of data and access to various data sources, and the ability to analyze it in various ways and at various levels for real time, historical, present, and predictive insight.

Connecting any device to anything else means new potential access points exist to those devices for monitoring and control. It means the potential for small to very large amounts of data that may now exist on, and flow between, multiple points. While this creates advantages for analysis, it also imposes security concerns that can involve operations within a company all the way to privacy and legal implications depending on when, where, and how that data is used. An example of this is the legal considerations in the movement of data between countries.

With these considerations in mind, at its core, an IoT architecture may use the following technology areas to define and address the needs described earlier. While this is a more general view of IoT application needs, not all elements may be used in every IoT implementation. Their use depends on the scope of the IoT application itself as described in the following bullet list:

- Device connection and data acquisition
- Data aggregation
- Edge analysis in or near the devices themselves
- Deeper Integration and analysis with IT systems and data
- Security considerations

Taken together, a high-level approach to IoT architecture can be represented as in Figure 20-2 that shows the physical and data integration that represents a more complete IoT approach. Additionally, software elements to provide device connection, access, data acquisition, security, analytics, and other functions within the IoT architecture are also necessary.

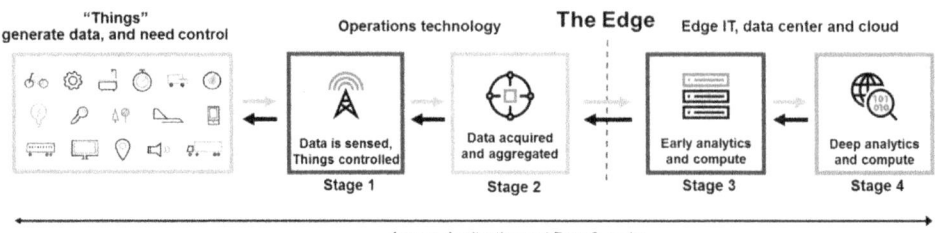

Figure 20-2 Core IoT Components

Projects focused strictly on operational benefit will focus more on the left side of "The Edge." Projects involving external data, cloud technologies, IT systems, and other similar integration points will require all elements.

Device connection and data acquisition (Stage 1)

To retrieve data from sensors and any other device, some form of physical connectivity, communications protocol, and software elements must be established between the data source and the point(s) accepting and aggregating data from that, and other, sources.

Choosing the right connection types and protocols depends on data locality, volume of data, and real time or mission critical nature of the project. For example, retrieving data and monitoring an airplane engine has limited locality (the airplane itself), a very large volume of data (large amounts of data per flight, per engine), and obvious mission critical need. Connecting the plane itself to data center IT systems or other data sources while in-flight has much broader geographical implications, less data moving outside the aircraft (at least while in-flight) compared with on-board data coming from local monitoring of the engines themselves, but with the same mission critical operation.

Connectivity to meet these many attachment needs can take on different types as shown in the following example list. There are many sources of information that provide detail, distance, and data transfer rate specifications, standards and specific connection methodologies for the methods below. These specifications are not discussed here.

- **Local connectivity** (for example, machines connected to local compute/control devices on a factory floor, sensors, actuators, Radio Frequency Identification (RFID), Digital Input/output, Power-over-Ethernet (POE), Small Computer System Interface (SCSI), RS-232, Bluetooth, video/audio connections)

- **Limited geographical connections** (for example, Wired or Wireless LAN connections, Cellular connect within a "cell," Fiber Optic connections)

- **Broader connectivity** (for example, general Internet connectivity, Satellite links, Microwave, Broader cellular networks, Radio, Low-power Wide-Area Network (LPWAN))

The goal of all these options with respect to device connectivity is to provide the appropriate level of access and data volume movement, and possibly control, in either one direction or bidirectionally as shown in Figure 20-3.

CHAPTER 20
The Power of Connection: The Internet of Things

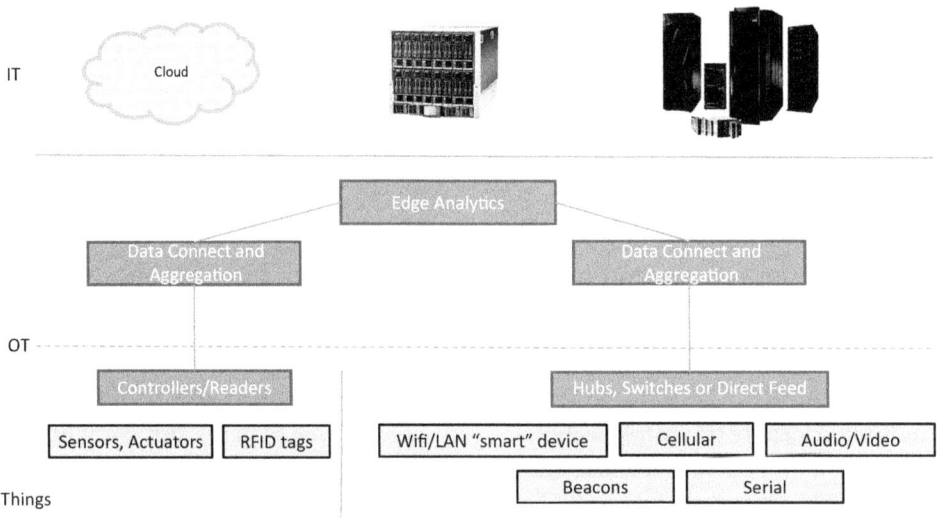

Figure 20-3 Examples of IoT core connectivity

Data aggregation (Stage 2)

Depending on the number of devices or objects connected, the volume of data each generates, and how quickly that data needs to be used for analysis, the data collected may either need to be left in raw form, or summarized before being passed along to other systems.

Real-time systems, such as a control valve on a pump, are those that require raw data to be collected in volume, analyzed, and immediate control or feedback decisions to be made. Most of these systems tend to focus on industrial applications such as energy, manufacturing, transportation, or military needs where decisions need to be made quickly for efficiency, safety, or other operational reasons.

Other industries, such as retail, agriculture, and medical systems may use a real-time approach for some specific application areas, such as hospital room monitoring, but generally need raw machine data to be summarized and decisions made in a historical data context such as a holistic view of patient care.

Figure 20-4 shows a summary of this data collection and aggregation approach.

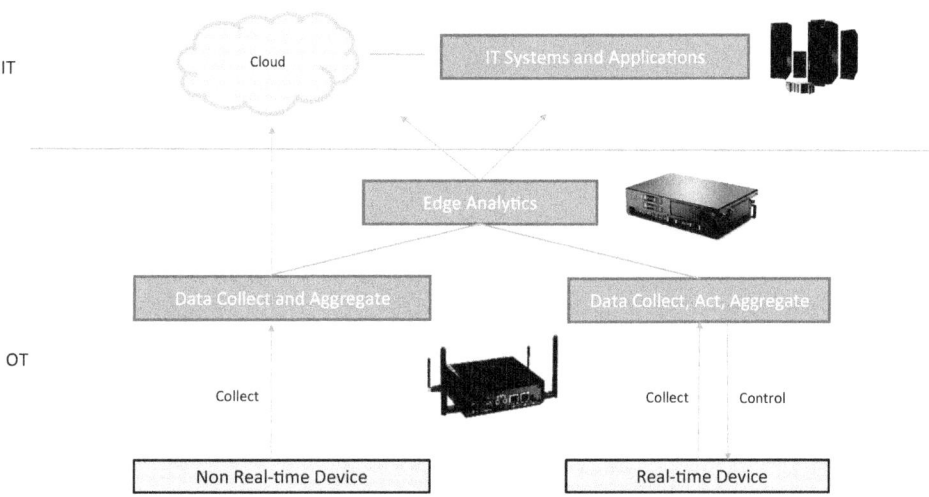

Figure 20-4 IoT Data collection and aggregation

Depending on the volume and timeliness of analysis, data can require aggregation at one or more levels. In some applications, only the data needed to control the device or provide analysis may be used, with only summary data passed up to other edge analytics and IT/cloud levels for deep analytics. In other applications, all the data may eventually be required for different levels of analysis. This means data may be stored at different levels, such as databases that reside at the edge computing level, or within cloud/IT systems that may store the raw or aggregated data.

In our aircraft example, the large volume of data generated by the engines require monitoring and control on-board during flight. This high volume of data can be collected and analyzed at the edge; on the aircraft itself. However, transmitting massive amounts of raw engine and other data to ground-based IT systems or cloud storage would be prohibitive as it is generated and cannot be guaranteed in terms of uninterrupted communications while in flight. Either summary aggregated data could be sent to those systems during flight operations, or all raw data could be sent for storage and analysis once the aircraft has completed its scheduled flight and at the gate.

This data collection and aggregation can be accomplished in or near the devices themselves by using technology such as HPE Edgeline products and software products that provide device connectivity and summarization/analysis of data for every type of vertical market application.

Edge analysis (Stage 3)

Once devices are connected and data is collected from them, the data can then be used in raw or summarized form as described earlier. In the case of real-time applications, this raw or summary data is generally used to monitor and control devices and provide input for broader system views for multiple devices being aggregated. In nonreal-time applications, generated raw or summary connected device data may be targeted primarily toward predictive or historical analysis, integrated with other data sources.

In traditional IT models to date, analytics has generally resided in the domain of IT systems, or in the operational realm, for very specific operational functions on systems such as Programmable Logic Controllers (PLCs.) and Remote Terminal Units (RTUs). With the advent of IoT, computing power for analysis is possible at the "edge"; closer to the objects generating the data themselves.

Edge computing involves data capture, aggregation, and analysis can be done within a connected device, such as an EKG heart monitor, or provided as a local service to a system of devices such as in an automobile. It may not be necessary to send all data to computational systems that reside in a data center or within a cloud service. While appropriate for some applications, the data analysis can often be done where the devices exist and where the data are generated, whether that is a retail store, locomotive, manufacturing floor, or oil rig, as examples. Figure 20-5 depicts the various analytics options.

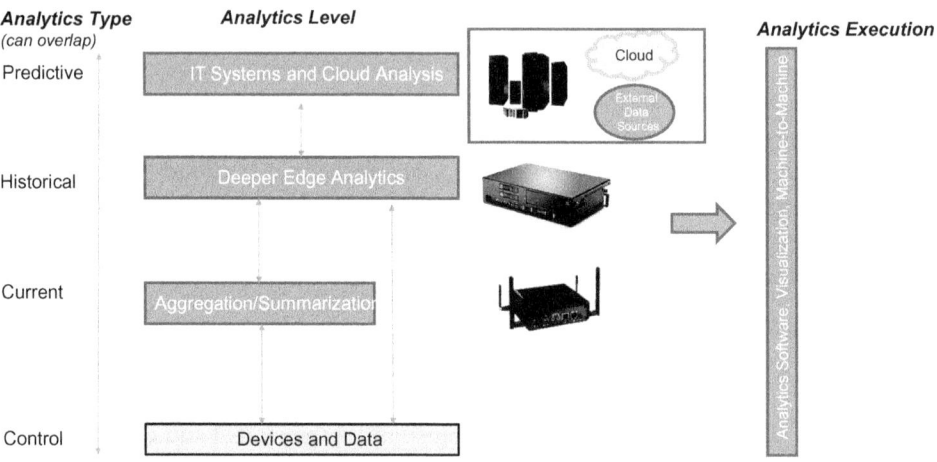

Figure 20-5 IoT Analytics Levels

In the world of IoT, the combination of more pervasive computing power at the edge, coupled with access to data from device sources, internal IT and external sources, can allow at least several different types of analysis to occur as described in the list below:

- **Control**—Analysis used to determine, state, and compare to an expected result for controlling one or more connected devices or systems.
- **Current**—Targeted at "what is happening now." May or may not have a control aspect to it. May use raw data from one or more devices or summary data to make these determinations.
- **Historical**—Analyzes of data from a single device, multiple devices in a system, or multiple systems. This view focuses on "what happened" to draw inference on control and operational impact.

- **Predictive**—Utilizes data from all the above sources, either raw or summary, to look at what *is* happening and what *did* happen so that determination can be made on what *will* happen in the future. This requires combining data from multiple device and other sources that complement the analysis with factors that help make the analysis more deterministic.

To this point, most companies have focused operationally on control and current methods of analytics with respect to connected devices. Traditional databases and IT systems have provided a platform for historical analysis and reporting on past events.

In moving analytics processing power and software to the edge, coupled with access to sources of data from those devices, IT and other external sources, the goal of more real time, and both historical and predictive analytics is possible on a much broader scale.

Deep integration with IT systems and other data sources (Stage 4)

While much of the focus of IoT is on device connectivity, data collection, and analytics at the edge, great potential exists for the integration of data from not only those elements but also traditional IT sources and external sources such as social media and external data services related to the company business focus. By combining device and system operational data with internal IT systems that maintain historical and business information, both the IT and operational elements in a company can benefit from improved operational analysis and business insight.

As an example, consider building maintenance, including heating and cooling systems, elevators, lighting, among others. These systems can be monitored and controlled in real time, viewed from the perspective of current and historical performance, repairs, and efficiency of use.

But consider adding to these elements, data from IT on performance of multiple like systems and devices from other locations, system vendor performance information on their products, weather data, building access statistics, electricity consumption information, repair records from the broader population of like systems. Properly analyzed, the resulting business and operational insights gained from this integration might save cost in utilization, detect trends, and predict repair needs prior to malfunctions taking place. It also may provide competitive advantage to the company maintaining the buildings and be offered as a monetized service to their customers.

The integration of IT and operational data elements is an area with a potential for significant value, a platform for achieving a truly predictive analytics approach with respect to IoT, and an important consideration in determining objectives for more advanced IoT projects.

Security considerations (supporting all stages)

One of the main concerns that companies have in moving toward IoT is in the area of security. This is a valid concern considering that even traditional IT systems experience security exposure. Adding to this the fact that IoT-connected devices expand the potential security risk points, and broaden out the security perimeter to wherever data and devices are accessible (essentially the world as a whole), security must be an important consideration across all elements of IoT architecture.

Security in the IoT sense must be thought of like other IT applications from several perspectives as described in the list below:

- Device access and policy
- Applications
- Data transit
- Data storage

With respect to device access, connecting various devices and objects also means they, and their data, are subjected to access. Technologies such as HPE Aruba ClearPass exist to create access policy management definitions that limit access to connected devices to people and systems approved to do so.

Applications are a primary source of security concern, even in the IT world. Application scanners can be utilized to determine security "holes" in applications and fix them, minimizing the potential for breaches through an application window.

Since a major part of IoT architecture also revolves around data collection, storage, movement, aggregation, and analysis, it is important to provide protection for all data, not just where and how it is stored, but as it moves between levels of IoT. Technologies exists for encrypting data within storage, such as files and databases, and as it moves across systems.

Together, a view of security that addresses devices, data, and applications will help ensure a reliable IoT project.

The potential of a connected future

The rapid emergence of new enabling IoT technologies is already showing great potential to improve operational efficiencies and create more advanced analytical models for current, historical, and predictive views of operations. These technologies are also giving rise to new avenues for rethinking current business models for additional insight and value.

Many early implementations of IoT projects have focused on operational efficiency, connecting more devices throughout a business, gathering data, and providing more advanced analysis to execute better from an operational perspective. Some industries, such as transportation, are advancing quickly past this evolution point, using IoT to create new consumer and business competitive value in addition to operational improvement.

From this initial operational view of IoT, Figure 20-6 shows a potential evolution of IoT in the future.

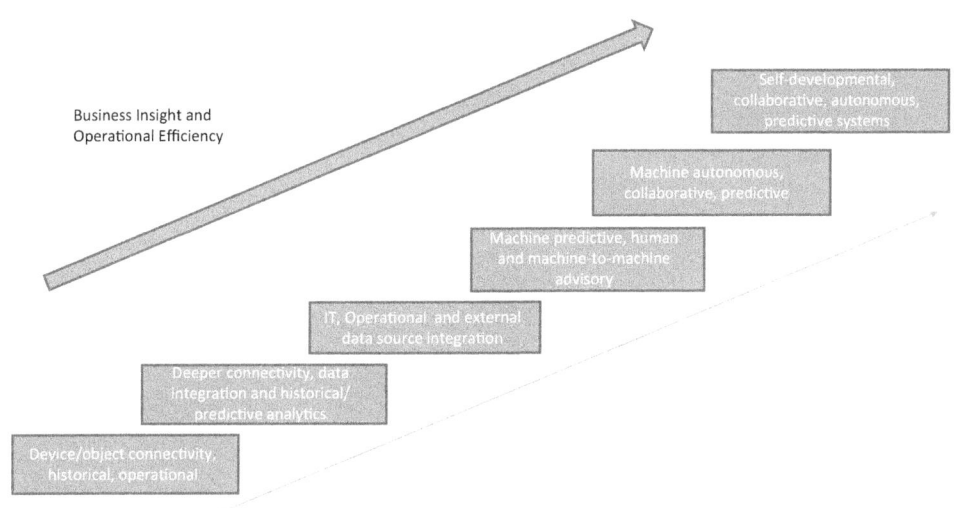

Figure 20-6 IoT possible evolution

Today's device and object connectivity and historical analytical view of data may evolve over time to add to many existing real-time IoT systems (for example, power distribution, robotics, PLCs) and incorporate additional sources of information within and outside a company. The value of this additional data, as described earlier for IT and operations, will be to improve both business insight and operational execution in a predictive sense, using a more holistic view of relevant data across the enterprise and other available sources, not just among the devices themselves.

From this point, IoT may evolve to systems such as computational systems, business and operation systems, and devices, helping to make predictive determinations and detecting patterns and trends that would be valuable in an advisory role to humans, such as a farmer in a connected tractor, or machine systems, such as supply chain systems, inventory, sales systems, that can plan and advise based on historical, current, and external data sources. This can also be advanced by continued technological innovation, standards development, and open platforms.

The move toward autonomous and collaborative systems would require a foundation in more pervasive data access, the further development of deep machine learning and artificial intelligence, plus agreed-upon collaborative communication protocols for machine-to-machine communications in real time. Such systems might even eventually be able to develop their own approach to problem-solving, given the results of data analysis.

As an example of autonomous operations, consider connected objects that can communicate with one another, sharing data on conditions and status, and making adjustments autonomously in synchronized agreement as they operate. These connected objects could not only advise one another but also inform other external systems that may have impact on their successful operations.

While some industries are midpoint and above on this evolution, most companies are just beginning their journey to IoT. All have the potential to advance their business. It is an exciting time in the computing world to see this potential just beginning to emerge.

21 Memory-Driven Compute and Composable Infrastructure

THE FUTURE: SHORT AND LONG TERM

The next few years have the potential to change the way that we view the infrastructure to run your workloads. Here are two observations about future technology:

- Composable Infrastructure, which is available today and called HPE Synergy, is a fluid pool of IT resources. You can quickly allocate resources for a given workload, expand the resources for that workload on-the-fly, and then reclaim the resources when you are done with the workload. The overall resource pool can be expanded by simply connecting the additional resources of any type including server, storage, and networking. The management of these resources is done using a simple, yet highly functional, management interface.

- Over the next few years, a completely new technology will emerge that will result in a dramatic change in the way we run workloads with Memory-Driven Computing with a proof point being "the Machine." The building blocks of the Machine will be radically different than what we have been accustomed to for decades. Networking will be accomplished with photonics, storage will be vast pools of nonvolatile memory, and compute will be performed by system-on-a-chip that can be optimized for each workload.

This view of the future starts with the long-term vision of Memory-Driving Computing and the Machine. It then covers Synergy that is available today and will greatly increase in capability over the next few months and years. You will quickly see that Synergy has some of the advanced capabilities of the Machine incorporated into its design.

Memory-driven computing

This is a completely different section than any other in the book. HPE is developing a radical new approach to future computing needs that will support data-centric world we are going to experience in the next few years. This will result in a quantum leap in performance achieved with the following building blocks:

- **Universal Memory**—The key aspect of Memory-Driven Computing and the Machine as a proof point will have main memory and persistent storage combined into a vast amount of storage that will have minimal latency and greatly reduced power consumption. The following are some of the key features of the memory components of which Universal Memory comprised that will have a big impact to support a new and advanced way of computing:
  - Memory is nonvolatile that means if there is a power loss, the memory maintains its contents.
  - Memory is highly dense that means a vast amount of storage will be contained in a small area.
  - This memory is fast and operates at the speed of system memory.
  - Minimal power is required to store vast amounts of data that is less than the power consumed by a flash drive
- **Photonics and the memory fabric**—Photonics will replace copper-based electrical connections. These optical communication links will be implemented by using micrometer-scale lasers on microchips to convert electrical signals to light. The light will then be converted back to electrical signals using the same microchips. This will result in greatly reduced latency. Today's networking and storage communication stacks will have to be modified in order to support this lightning-fast communication method.
- **System-on-chip (SoC) processing**—We will move from general purpose processing to SoC that will result in both higher compute performance and greatly reduced energy consumption.
- **Operating system**—There is unlimited opportunity for a new operating system to take advantage of the advancements in Memory-Driven Compute and the Machine such as vast amounts of universal memory. Traditional operating system techniques, such as swapping, will not be required in this environment. First, the operating system will allow legacy applications to run in this advanced environment and then, in phase two, developers will be able to take full advantage of the full capabilities of the Machine and achieve orders-of-magnitude improvements in speed and scale.
- **Additional considerations**—Under development are the following technologies related to Memory-Driven Computing:
  - **Security**—Higher levels of security, in both architecture and operating system, are required with so much data in Universal Memory that is subject to cyber-attack.
  - **Management**—The Machine will have no boundaries in terms of nodes interacting with one another. Management of countless nodes requires a new management technique to configure and maintain nodes in this distributed environment.
  - **Analytics**—In this highly distributed environment, new analytics techniques are being developed to handle multipetabyte datasets on which the Machine will manipulate.

o **System architecture**—Universal Memory, SoC, and photonics will require a new system architecture optimized for these advanced technologies to meet the unique capacity and performance that will be part of the Machine.

These components result in a new approach to computing that is memory-driven rather than our traditional processor-centric as depicted in Figure 21-1.

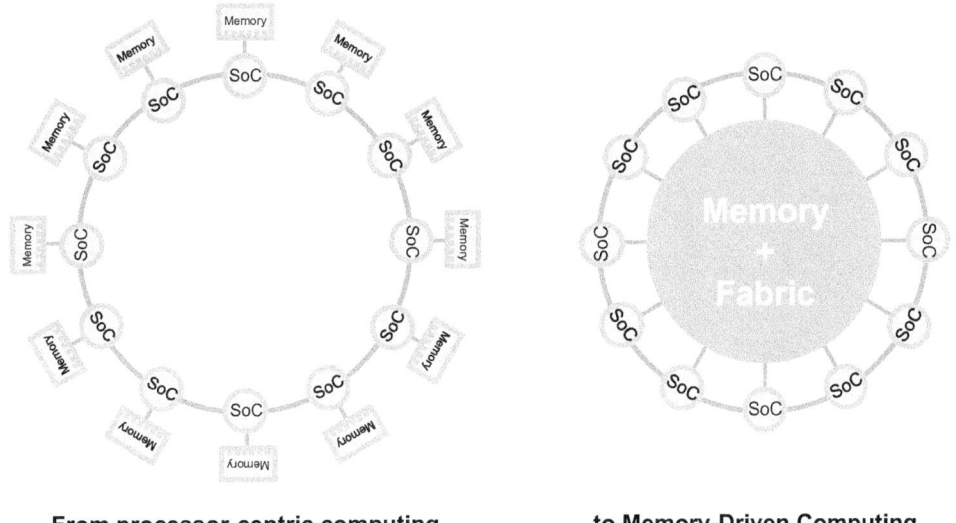

From processor-centric computing…   **…to Memory-Driven Computing**
-All copper replaced with Photonics

Figure 21-1 Depiction of memory-driven computing

Each SoC will have private memory, used by only that individual SoC, and then the Universal Memory, that will be shared across all of the racks.

This massive amount of shared memory is shown in Figure 21-2.

CHAPTER 21
Memory-Driven Compute and Composable Infrastructure

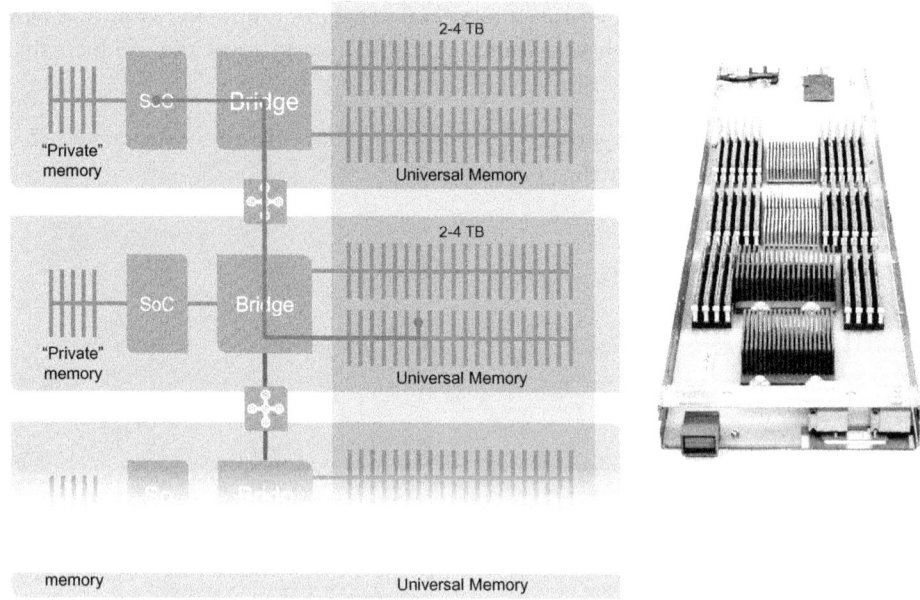

Figure 21-2 The Machine and Universal Memory

You can imagine over 100 racks of the Machine with this architecture, supporting hundreds of Peta-Bytes of Universal Memory that will change the way that large problems are solved. This will result in a different way of programming altogether. A microservice, for instance, may run in one location of Universal Memory and be accessible to the Machine across multiple racks.

The memory-driven approach will result in performance that will skyrocket and power consumption that will drop dramatically. One of the key reasons for the performance increase is that huge pools of Universal Memory will obviate the need to access much slower storage. The power consumed for operations will have a commensurate power reduction providing two huge advantages to universal memory as shown in Figure 21-3.

Level	Amount	Access time (approximate, in processor cycles)
Register	$O(2^6)$	1
L1 cache	$O(2^{15})$	2
L2 cache	$O(2^{18})$	10
L3 cache	$O(2^{22})$	60
Memory	$O(2^{37})$	300
SSDs	$O(2^{39})$	250,000
Disk	$O(2^{41})$	7,500,000

Operation	Energy (pJ)
64-bit integer operation	1
64-bit floating-point operation	20
256 bit on-die SRAM access	50
256 bit bus transfer (short)	26
256 bit bus transfer (1/2 die)	256
Off-die link (efficient)	500
256 bit bus transfer (across die)	1,000
DRAM read/write (512 bits)	16,000
HDD read/write (32k bits)	$O(10^6)$

Figure 21-3 Reduced Clock Cycles and Power Consumption

The right side of Figure 21-3 shows that accessing a register, which is ideal, is only one processor clock cycle versus a disk access with is 7.5 M clock cycles. Similarly, a 64-bit operation consumes one pico-joule versus a disk access that consumes 10^6 pico-joules.

The left side of Figure 21-3 shows the capacity of each level of storage with theta under "Amount" and the access time in terms of processor cycles. Reducing the number of processor cycles to access stored information, in some cases dramatically, along with the dramatic power reduction make this a technology that, when widely available, will be embraced by our industry.

The photonics in the Machine will consist of three levels: within the Machine frame, between frames, and between racks. This means that communication at all levels will take place with photonics.

The following URL provides some musings on the types of applications that would benefit from this new and advanced computing technique as well as some of the experts involved in the development of the Machine: http://www.labs.hpe.com/research/themachine/

HPE Synergy: Available Today

HPE Synergy, which is available today, is a fluid pool of IT resources that you can use as you see fit. The key benefits are operational velocity, "frictionless IT," and reduced costs. With this new concept, you will be able to realize these benefits as described below:

- With "frictionless IT," you will be able to spin-up new services almost instantly thereby becoming a Service Provider to your organization. In the new DevOps world, this is key to being efficient and competitive. The long planning, procurement, installation, and other processes are obviated by Synergy that is waiting for you to deliver new services.

- Costs are greatly reduced with the immense amount of time and expertise saved by using Synergy. Resources can be better spent on bringing value to your business in the form of strategic projects such as new applications that can benefit your business.

- With operational velocity, intelligence and automation are built into Synergy that is software-defined from the ground up. For example, with only one line of code can abstract every element of Synergy which is also referred to as "infrastructure as code." The following example shows the format that is used for infrastructure as code:

New-HPOVProfile -name myCloud -template SynergyCloud

To some organizations, these benefits are dramatic. Spending much less time managing an environment, and instead working on more business beneficial endeavors, can make a world of difference.

From a component standpoint, Synergy consists of the following:

- HPE Synergy 12000 is the frame that accommodates compute, storage, fabric, and management. The frame is designed to accommodate many generations of these components to ensure investment protection (Table 21-1).

Table 21-1 HPE Synergy 12000 Frame specifications

Model	12000 Frame
Race units	10U
Compute bays	12 half-height, 6 full-height
Module types	Half-height, full-height, double-wide full-height compute modules, double-wide half-height storage module
Fabrics supported	3+3 Redundant Fabric Modules Ethernet, Fibre Channel
Management	HPE Synergy composer powered by HPE One View
Midplane bandwidth	15 Tbps
Cooling	10 fans (included)
Power	6x 2650 W, 96% efficiency, –48 V dc, 277 V ac, 380 V dc

- Synergy Compute modules have a variety of capabilities and evolve with processor technology. At the time of this writing, there are numerous compute modules that go up to four processors

- Synergy Storage has a variety of storage tiers and density options. The following list shows the three types of storage that can be deployed:

    o Direct Attached Storage (DAS) modules to provide a large pool of storage.

    o Software Defined Storage (SDS) that take DAS and turn it into shared storage.

    o Storage Attached Network (SAN) that can be shared among Synergy frames.

- Power and Cooling in the Synergy frame is designed for the future providing enough power and airflow capacity to support high-end processors and high-memory capacity.

- Synergy Composer is an appliance in the frame that enables management of compute, storage, and fabric using OneView. Composer integrates into the frame and can manage from one frame or multiple racks. Multiple racks are connected through a dedicate 10 Gb interface called Frame Link Modules (FLM.) The FLM uses industry best practices to air-gap management and production networks, and provides all information about the frame including device inventory, power, and thermal data to the Composer.

- Synergy Streamer provides compute nodes with bootable images. "Golden images" are stored on the streamer and can be quickly deployed to compute modules in the frame. Streamer simplifies lifecycle management by providing a simple way to update Golden Images and rapidly rolling out new golden images to existing compute modules.

- Synergy Fabric employs a Master and Satellite topology, providing high-speed communication across multiple frames, while reducing CAPEX and OPEX as multiple frames can be deployed and managed as a logical entity without adding additional management overhead or latency.

These components make up the Synergy solution shown in the Figure 21-4.

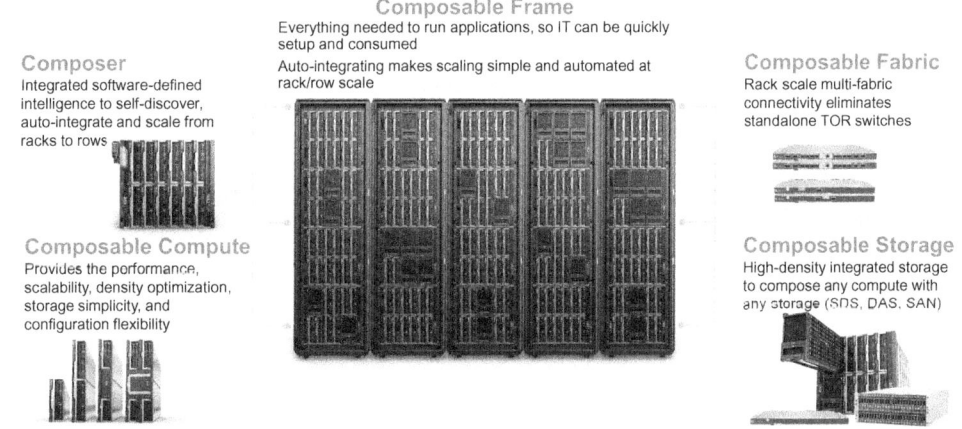

Figure 21-4 HPE synergy rack solution

Figure 21-4 shows the components as they would be configured into a Synergy rack environment.

The next section covers Synergy management that is one of the key aspects of the composable solution.

Management of Synergy

In order to have a fluid pool of IT resources, you need a management environment that will scale from one frame to many. The management environment needs to view, configure, manage, and perform many other tasks in order to achieve the fluid pool of IT resources. Some of the components that are part of integrated management environment are mentioned in the following list:

- The Composer is an appliance that can manage multiple Synergy frames and a key component of the fluid pool of IT resources. This appliance deploys, monitors, and updates the infrastructure from a single location. This includes compute, storage, and fabric.

- The Frame Link Module (FLM) is the control point and link to multiple frames. The FLM employs a dedicated 10 Gb Ethernet network. This fabric uses air gap isolation as a security measure. The FLM contains asset and inventory information for its frame including thermal and real-time power usage of the frame.

- HPE OneView provides RESTful API to manage Synergy components. This provides a programmatic access to Synergy using PowerShell and Python. OneView is embedded in the Composer.

- Image Streamer provisions Computer Modules with bootable golden images. Using server profiles and templates images are streamed directly to Compute Modules. You can configure, provision, and update infrastructure with templates that are ideally suited to a specific workload.

An example showing how the Composer and Image Streamer work together to deploy a stateless compute module, which is one that does not need to retain any hardware state, is shown in Figure 21-5.

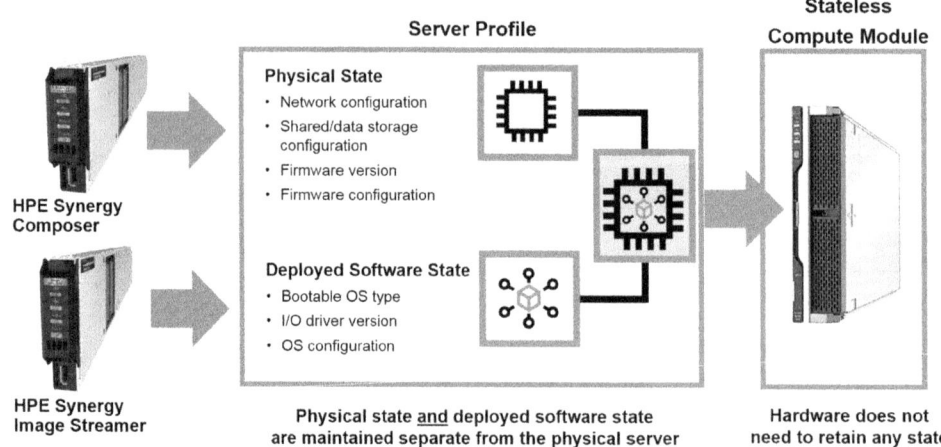

Figure 21-5 Composer and Image Streamer working together

Figure 21-5 depicts a template and operating systems image with all relevant physical attributes, such as network configuration and firmware definitions, being deployed to a compute module. The new image will be ready to boot in seconds. Many operating systems and hypervisors can be deployed using this technique.

Epilogue

Thank you for taking the time to review this material and congratulations on making it to the end of the book! My objective now is to solidify what you have learned, give some indications as to next steps, and leave you with a few parting thoughts to guide you on your journey.

As I mentioned at the outset, workloads are key to practically every technology decision. If you are a technology decision maker or influence how technical buying decisions get made, then you've probably asked this question more than once: how can I best take advantage of the information in this book? My hope is that real-world workload examples featured throughout have resulted in more than a few light bulb moments of improvements to be made and designs to be considered. Since every workload solution is a response to a unique set of challenges and conditions, there's a better than even chance that the solutions presented may have fired up a lot more questions as opposed to delivering a neatly packaged set of answers. To which I say, great, you are exactly where you need to be!

My extensive experience in Hewlett Packard Enterprise's New York City's Customer Experience Center (CEC) and working with many customers has taught me that with HPE's extensive product line, technical experts are looking for specific recommendations on the way to solve their problem and integrate it into their existing infrastructure. This may be true for you as well. Hadoop, for instance, can be implemented in many different ways and so what our customers want to see is a definite design proposal, such as HPE's Big Data Reference Architecture (BDRA). The BDRA shows why HPE's Apollo servers, separating compute and storage, provides a better design choice for this use case to the HPE ProLiant DL380 Gen9 servers. In short, IT professionals want to know the ideal way of solving the problem and this book provides some good illustrations of how to drive that outcome.

Since there are multiple ways to solve every workload problem, this book is not meant to be prescriptive. In fact, I'd advise you to avoid anyone that tries to convince you that designing and implementing efficient and effective workloads is simply a matter of plug and play. To quote a colleague from a preceding chapter: "As with many complex solutions, problems will inevitably be encountered and so the right set of experts are needed to ensure mission critical applications get up and running quickly." Even more than products and technologies, the secret to successful IT optimization is in making the right series of choices to navigate the challenges you face and to find a partner with the technical expertise to make things happen. It is desirable to look beyond just a quick fix and be well positioned to anticipate and respond to future trends. As indicated in the chapters highlighting hybrid IT, the shift to multi-cloud environments, and industrial IoT at the intelligent edge, HPE is aligned with open and flexible solutions that accelerate your IT transformation, now and for the future.

I work with customers every day who have workload challenges and want to engage my team to explore the alternatives, thereby landing on the best solution for their environment. My primary goal

Epilogue

in this book was to show the way in which my technical experts solved specific workload problems in specific environments. Every one of these challenges had more than one potential solution and, after carefully considering alternatives, we were able to implement the ideal solution for their environment.

I wish you every success in taking the next leap forward on your own IT journey! And if you need any recommendations on next steps for your particular needs, feel free to contact me at marty.poniatowski@hpe.com. I'd be more than happy to hear from you or help point you in the right direction.

I'm very interested in your comments on workloads you'd like to see covered in the next revision of this book, so please share your thoughts with me.

Index

A

Account Support Manager (ASM) 200
Active Directory (AD) trusts 103
Adaptive Optimization effect 184
Advanced Power Manager (APM) 30, 40
Advanced tests 38–39
Age of Algorithmic Businesses 7
Americas Integration Center (AIC) 206
Apollo Platform Manager (APM) 12, 17
Apollo servers 198
  Apollo a6000 chassis 18
  Apollo-based solution 29–31
  Apollo 2000 chassis 196
  Apollo 4200 servers 196
  Apollo 4530 servers 60
  Apollo 6000 chassis 12
  Apollo 6000 Power Shelf 18
Application Programming Interface (API) 210
Archiving large files
  disk-based storage environment
    archive solution 58–59
    controllers 61–64
    crafting 57
    hardware inventory 64–65
    hardware RAID settings 61
    implementation 66
    server and storage 59–60
    solution validation and success criteria 66–67
  object storage
    problem statement 67–68
    RING 68
    solution 69–71

B

Balanced and density optimized (BDO) 193
Baseline tests 38
Basic Input/Output System (BIOS) modifications 40
Bill of Materials (BOMs)
  archiving large files 64–65
  BLc7000 Blade Infrastructure 99–100
  Citrix XenApp 7.6 environment 107–108
  connector servers 47–48
  CS900 configuration 142–143
  Disaster Recovery 186
  EMR Software 121–125
  eVDI 99–100
  Hadoop architecture 201–205
  HC250 and backup 171–175
  HC380 VDI solution 86–88
  HPC 17
  hybrid IT 217–220
  Mellanox Switches 27
  scale-out solution 33–34
  single DL380 Gen9 Management Servers 27
  single GPU 27
  single High Memory 26

single Standard Memory 26
SQL Server workload solution 149–150
storage servers 46–47
supervisor server 48
virtual machines 161–163

BladeSystem 116–117

BLc7000 Blade Infrastructure
benefits 89–90
customer requirements 97–98
datacenter 95–96
DR site 99
hardware components 93–95
high-level overview 91
network and SAN connectivity 97
NVIDIA GPU integration 98
processor selection 99
requirements 90–91
scale-out rack view 92–93
solution inventory
   BOM 99–100
   software components 100
3Par Storage solution 97
user demographics 96

C

Citrix 110
benefits 101
BOM 107–108
C7000 BladeSystem and 3PAR 104
Data Store 106
Desktop Delivery Controller 106
Director Servers 107
Environment 106
hardware requirements 109
HDX technologies 96
high-level components 102

License server 107
Microsoft licensing requirements 109
multi-location configuration 103–104
NetScaler 103–104, 107
Provisioning Services 106
rack diagram 102–103
software requirements 109
Storefront 106
StoreVirtual technology 105
Studio management 106

Citrix Xen Virtual Delivery Agent (VDA) 7, 74

Cloud Architect training 221

Cloud models 3

Cloud Operator training 221

Cloud Service Automation (CSA) 213–214

CloudSystem 10, 212–213

Cluster Management Utility (CMU) 40, 200

Composable Infrastructure
advantages 239–240
components 240–241
management of 242

Configure-To-Order (CTO) 129

ConvergedSystem 700 (CS700)
characteristics 111–112
ConvergedSystem 2.0 solution 113–116
customer environments 117
deduplication 118
deployment 116–117
hardware inventory 121–125
management and monitoring 120–121
ODP workload 118
RDBMS workload 118
snapshots and clones 118
software inventory 121–125
3PAR Four-Node 7440 AFA 119–120

Converged System 900 (CS900) 137, 140, 141
Custom Scripts 40

D

D0620 enclosures 63–64
Data acquisition 227–228
Data aggregation 228–229
Data collection 228–229
Data Entry Identification Document (DEID) report 40
Deployment
    CS700 116–117
    Disaster Recovery 190
    Hadoop 195
    HPC 19–20
    SAP HANA 137–138, 144
    Superdome X 148–149
Device connection 227–228
Direct Attached Storage (DAS) 240
Disaster Recovery (DR) 110
    Adaptive Optimization effect 184
    benefits 101
    BOM 107–108
    C7000 BladeSystem and 3PAR 104
    Citrix Farm 105–107
    configurations 180–181
    deployment 190
    hardware inventory 109, 186–189
    high-level components 102
    layouts and specifications 184–185
    Microsoft licensing requirements 109
    multi-location configuration 103–104
    performance and capacity utilization 183–184
    rack diagram 102–103
    servers 182–183
    site 99
    software inventory 109, 186
    storage solution 182
    StoreVirtual technology 105
    targets 186
    validation 190
DL360 Gen 9 servers 196–198
DL380 line 46
DL360 servers 47–48

E

Edge analysis 229–231
Electronic Medical Records (EMR) software
    characteristics 111–112
    ConvergedSystem 2.0 solution 113–116
    customer environments 117
    deduplication 118
    deployment 116–117
    hardware and software inventory 121–125
    management and monitoring 120–121
    ODP workload 118
    RDBMS workload 118
    snapshots and clones 118
    3PAR Four-Node 7440 AFA 119–120
Engineering Virtual Desktop Infrastructure (eVDI)
    benefits 89–90
    customer requirements 97–98
    datacenter 95–96
    DR site 99
    hardware components 93–95
    high-level overview 91
    network and SAN connectivity 97
    NVIDIA GPU integration 98
    processor selection 99
    requirements 90–91
    scale-out rack view 92–93
    solution inventory

Index

BOM 99–100
   software components 100
3Par Storage solution 97
user demographics 96

Enhanced Data Rate (EDR) 22

Epic solution
   characteristics 111–112
   ConvergedSystem 2.0 solution 113–116
   customer environments 117
   deduplication 118
   deployment 116–117
   hardware and software inventory 121–125
   management and monitoring 120–121
   ODP workload 118
   RDBMS workload 118
   3PAR Four-Node 7440 AFA 119–120

Extract Transform Load (ETL) 118, 119

F

Factory Express 39
Factory Integration Services 133
Fiber channel instructions 164–165
Fiber Channel Over Ethernet (FCOE) 13
Fiber Channel SAN infrastructure 149
Financial Services Industry (FSI) 73
Frame Link Module (FLM) 241, 242

G

Gigabit Ethernet (Gbe) 157

H

Hadoop
   Apollo 2000 chassis 196
   Apollo 4200 servers 196
   compute node design 199
   design factors
      cost 195
      deployment 195
      flexibility 194
      floor space 194
      growth 193
      performance 193–194
      power 194
   DL360 Gen 9 servers 196–197
   hardware inventory 201–205
   high-level cluster overview 195
   implementation plan 206
   management node design 200
   network setup 199–200
   rack diagram 197
   selection
      compute platform 198–199
      drives 199
      management tool 200
      proactive care advanced support 200–201

Hadoop Distributed File System (HDFS) 200

HANA
   components 140–141
   deployment 137–138, 140, 144
   hardware inventory 142–143
   infrastructure design assumptions 140
   input sizing 139
   sizing and architecting process 138

Hard Disk Drives (HDD) 119–120

Hardware inventory
   advantages 129
   BLc7000 Blade Infrastructure 99–100
   business intelligence solution 16–19
   Citrix XenApp 7.6 environment 109
   compute nodes 132
   ConvergedSystem 700 121–125

CS900 configuration 142–143
Disaster Recovery 186
disk-based storage environment 64–65
Epic solution 121–125
eVDI 99–100
Hadoop architecture 201–205
HC250 and backup 171–175
HPE HC 250 components 127, 128–129
HPE HC 250 conceptual design 128
hybrid IT 217–220
individual components 129
Moonshot 1500 chassis 77–78
scientific program 25–28
SQL Server workload solution 149–150
TOR switch 128
virtual machines 161–163

Hardware power and cooling requirements 133

Helion CloudSystem 10 Enterprise 212–213

Helion Stackato 216–217

High availability (HA) 57, 82

High-performance computing (HPC)
business intelligence solution
  deployment 19–20
  hardware inventory 16–19
  PoC 14–16
  software inventory 19
  solution design 12–13
  technical team 11
scientific program
  constraints 21–22
  hardware components 22–23
  hardware inventory 25–28
  project plan 28
  server cooling 25
  server power 24–25
  servers and network infrastructure 22

server space 24
solution validation and success criteria 28

Hosted Desktop Infrastructure (HDI)
applications 75
Cartridge-to-switch connections 74
components 75
customer requirements 73
high-performance capacities 73
iLO 74
low-power processors 73
M710 family 76
solution design 76–77
solution inventory
  cartridge configuration 78–79
  characteristics 77
  Citrix components 79
  client configuration 78
  hardware components 77–78
  Moonshot power supplies 80
  Moonshot software 79
  system management 80
workload-optimized compute engines 73–74

HPE Apollo a6000 chassis 18

HPE Apollo-based solution 29–31

HPE Apollo 6000 chassis 12

HPE Apollo 6000 Power Shelf 18

HPE Apollo Platform Manager (APM) 12, 20

HPE Apollo power 12

HPE Apollo Power Manager 17

HPE Application Transformation services 220–221

HPE Bill of Materials (BOMs)
archiving large files 64–65
Citrix XenApp 7.6 environment 107–108
connector servers 47–48
CS900 configuration 142–143

Disaster Recovery 186
EMR Software 121–125
eVDI 99–100
Hadoop architecture 201–205
HC250 and backup 171–175
HC380 VDI solution 86–88
HPC 17
hybrid IT 217–220
Mellanox Switches 27
scale-out solution 33–34
single DL380 Gen9 Management Servers 27
single GPU 27
single High Memory 26
single Standard Memory 26
SQL Server workload solution 149–150
storage servers 46–47
supervisor server 48
virtual machines 161–163

HPE BladeSystem 116–117

HPE BLc7000 Blade Infrastructure

    benefits 89–90
    customer requirements 97–98
    datacenter 95–96
    DR site 99
    hardware components 93–95
    high-level overview 91
    network and SAN connectivity 97
    NVIDIA GPU integration 98
    processor selection 99
    requirements 90–91
    scale-out rack view 92–93
    solution inventory
        BOM 99–100
        software components 100
    3Par Storage solution 97
    user demographics 96

HPE CloudSystem 10 Enterprise 212–213

HPE Cluster Management Utility (CMU) 200

HPE ConvergedSystem 2.0 solution 113–116

HPE ConvergedSystem 700 (CS700)

    characteristics 111–112
    ConvergedSystem 2.0 solution 113–116
    customer environments 117
    deduplication 118
    deployment 116–117
    hardware and software inventory 121–125
    management and monitoring 120–121
    ODP workload 118
    RDBMS workload 118
    snapshots and clones 118
    3PAR Four-Node 7440 AFA 119–120

HPE Converged System 900 (CS900) 137, 140, 141

HPE DL380 line 46

HPE DL360 servers 47–48

HPE Epic solution

    characteristics 111–112
    ConvergedSystem 2.0 solution 113–116
    customer environments 117
    deduplication 118
    deployment 116–117
    hardware and software inventory 121–125
    management and monitoring 120–121
    ODP workload 118
    RDBMS workload 118
    snapshots and clones 118
    3PAR Four-Node 7440 AFA 119–120

HPE Flex Fabric 5700 switches 132

HPE HC 250

    design 128
    hardware inventory
        advantages 129

  components 128–129
  compute nodes 132
  network connection 132
  TOR switch 128
 project l implementation 133
 software inventory 132
 solution validation and success criteria
  high vailability 135
  MEM 136
  MPIO 136
  Network RAID-10 135, 136
  vCPU 134

HPE Helion Stackato 216–217

HPE Hybrid IT Education Services 221

HPE Hyper Converged 250 (HC250)
 benefits 101
 BOM 107–108
 C7000 BladeSystem and 3PAR 104
 Citrix Farm 105–107
 components 169–170
 goals 169
 hardware requirements 109
 high-level components 102
 implementation, deployment, and testing 176
 IOPs 170
 mega-bytes per second 170–171
 Microsoft licensing requirements 109
 multi-location configuration 103–104
 parameters 171
 rack diagram 102–103
 software inventory 175
 software requirements 109
 solution validation and criteria 176
 StoreOnce device 171
 StoreVirtual technology 105
 test workload results 176–177

HPE HyperConverged 380 (HC380)
 customer requirements 81
 hybrid IT 211–212
 processor selection 84–85
 solution inventory 86–88
 solution overview 81–83
 StoreVirtual VSA
  hypervisor requirements 85
  performance and capacity 85–86
 user activity assessment 83
 vCPU 83
 from Windows 7 to Windows 10 83

HPE Integrated Lights-Out (iLO) 66

HPE Integrity Superdome X 137

HPE Moonshot
 applications 75
 Cartridge-to-switch connections 74
 components 75
 customer requirements 73
 high-performance capacities 73
 iLO 74
 low-power processors 73
 M710 family 76
 Moonshot 1500 chassis 73–76
 solution design 76–77
 solution inventory
  cartridge configuration 78–79
  characteristics 77
  Citrix components 79
  client configuration 78
  hardware components 77–78
  power supplies 80
  software 79
  system management 80

HPE OneView 40, 104, 120–121, 242

HPE OneView InstantOn software 104

HPE Operations Analytics 114, 120
HPE Power Advisor 24–25
HPE Power Sizer Tool 133
HPE ProLiant DL380 Gen9 server 12, 113
HPE ProLiant XL230a Gen9 server 12, 14–15, 18
HPE's Americas Integration Center (AIC) 206
HPE SAP sizing and architecting process 138
HPE Scalable Object Storage
    connector servers 47–48
    deployment 51–52
    design 43–44
    erasure coding 49–50
    key-value store 44–46
    licenses 48
    scale-out 50–51
    sizing calculations 49
    storage servers 46–47
    supervisor server 48
    system management 51
HPE-Scality reference architecture 70
HPE Smart Array 12
HPE SmartArray controllers 66–67
HPE Smart Update Manager (SUM) 200
HPE StoreFront Analytics 114
HPE StoreOnce device 171
HPE StoreVirtual technology 105
HPE StoreVirtual VSA
    hypervisor requirements 85
    performance and capacity 85–86
HPE Superdome X platform 146–147
HPE Synergy
    advantages 239–240
    components 240–241
    management of 242
HPE Synergy 12000 Frame specifications 240
HPE Technical Services 133

HPE 3PAR 7440 117
HPE 3PAR StoreServ 7440c 4N storage solution 113
HPE 3PAR StoreServ software 114
HPE Virtualization Performance Viewer 114, 121
HPE workload and densityoptimized (WDO) servers 193
Hybrid IT
    advantages 209
    components
        CSA 213–214
        HC380 211–212
        Helion OpenStack 215–216
        Helion Stackato 216–217
        HPE CloudSystem 10 Enterprise 212–213
        Operations Orchestration 214–215
    education and application services 220–221
    hardware inventory 217–220
    IaaS 210
    requirements 209–210
Hyper Converged 250 (HC250)
    benefits 101
    BOM 107–108
    C7000 BladeSystem and 3PAR 104
    Citrix Farm 105–107
    components 169–170
    goals 169
    hardware requirements 109
    high-level components 102
    implementation, deployment, and testing 176
    IOPs 170
    mega-bytes per second 170–171
    Microsoft licensing requirements 109
    multi-location configuration 103–104
    parameters 171
    rack diagram 102–103

software inventory 175
software requirements 109
solution validation and criteria 176
StoreOnce device 171
StoreVirtual technology 105
test workload results 176–177
HyperConverged 380 (HC380)
    customer requirements 81
    hybrid IT 211–212
    processor selection 84–85
    solution inventory 86–88
    solution overview 81–83
    StoreVirtual VSA
        hypervisor requirements 85
        performance and capacity 85–86
    user activity assessment 83
    vCPU 83
    from Windows 7 to Windows 10 83
Hyper Converged Administrator training 221
Hyper Converged System
    hardware inventory
        advantages 129
        compute nodes 132
        HPE HC 250 components 128–129
        HPE HC 250 Network connection 132
        individual components 129
        TOR switch 128
    HPE HC 250 design 128
    project plan implementation 133
    software inventory 132
    solution validation and criteria
        high vailability 134
        MEM 136
        MPIO 136
        Network RAID-10 135, 136
        vCPU 134

I

Image Streamer 242
Infrastructure as a service (IaaS) 210
Input/Output operations per second (IOPs) 97, 170
In-rack Ethernet Cables 17
Insight Remote Support 40
Integrated Lights-Out (iLO) 40, 66, 74
Intel Xeon E7-8893 v2 6-core IvyBridge processor 148
Internet of Things (IoT) 6–7
    architecture elements 226–227
    data acquisition 227–228
    data aggregation 228–229
    data sources 231
    device connection 227–228
    edge analysis 229–231
    industrial perspective 223
    opportunity 224–225
    potential evolution 232–233
    security 231–232
Internet Protocol (IP) network 157
Internet Small Computer System Interface (iSCSI) 157, 165–167
Intl Xeon E7-v2 processors 148
IvyBridge processors 148–149

L

Local area network (LAN) 117

M

Mega-bytes per second 170–171
Memory-driven computing
    analytics 236
    clock cycles and power consumption 238–239
    depiction 237

Machine development 239
management 236
memory fabric 236
operating system 236
photonics 236
security 236
shared memory 237–238
SoC processing 236
system architecture 237
universal memory 236

Microsoft Remote Desktop Services (RDS) License Server 107

Moonshot
applications 75
Cartridge-to-switch connections 74
components 75
customer requirements 73
high-performance capacities 73
iLO 74
low-power processors 73
M710 family 76
Moonshot 1500 chassis 73–76
solution design 76–77
solution inventory
  cartridge configuration 78–79
  characteristics 77
  Citrix components 79
  client configuration 78
  hardware components 77–78
  power supplies 80
  software 79
  system management 80

Moonshot Component Pack (MCP) 79
Multi-cloud models 3
Multi-Path Extension Module (MEM) 136
Multi-Path IO (MPIO) 136

N

NetScaler 103–104
NetScaler MPX 5550 103
Network RAID-10 135, 136
Nonproduction environment 137–138
Non-Uniform Memory Access (NUMA)
business requirements 145
performance issue 153–155
technical requirements 146

NVIDIA® Tesla® GPU accelerator 90, 94

O

OneView 40, 104, 120–121
OneView InstantOn software 104
Open service catalog 209
OpenStack 215–216
Open System Interconnection (OSI) 213
Operating system 236
Operational Database Platform (ODB) workload 118
Operational Technology (OT) 6–7, 225
Operations Analytics 114, 120
Operations Orchestration (OO) 214–215

P

Power Advisor 24–25
Power Distribution Units (PDU) 17
Preboot Execution Environment (PXE) 79
Proactive Care Advanced support 200–201
Programmable Logic Controllers (PLCs) 230
ProLiant Apollo 4510 server 58
ProLiant DL360 Gen9 Server Management 113
ProLiant DL380 Gen9 server 12
ProLiant servers 200
ProLiant XL230a Gen9 Server 12, 14–15, 18
Proof of Concept (PoC) 12

R

RAID-10 135, 136
Recovery Point Objective (RPO) 160
Recovery Time Objective (RTO) 160
Redundant Array of Independent Disks (RAID) 61, 120
Relational database management system (RBDMS) workload 118
Reliability, Availability, and Serviceability (RAS) features 153
Remote Terminal Units (RTUs) 230

S

SAP Enterprise Resource Planning Software 138
SAP HANA
    components 140–141
    deployment 137–138, 144
    hardware inventory 142–143
    infrastructure design assumptions 140
    input sizing 139
    sizing and architecting process 138
SAP HANA sizing input 139
SAP infrastructure design assumptions 140
Scalable Object Storage
    connector servers 47–48
    deployment 51–52
    design 43–44
    erasure coding 49–50
    key-value store 44–46
    licenses 48
    scale-out 50–51
    sizing calculations 49
    storage servers 46–47
    supervisor server 48
    system management 51

Scalable storage
    two-site online object
        active/passive configuration 52–54
        requirements 52
    unstructured data storage
        connector servers 47–48
        deployment 51–52
        design 43–44
        erasure coding 49–50
        key-value store 44–46
        licenses 48
        scale-out 50–51
        sizing calculations 49
        storage servers 46–47
        supervisor server 48
        system management 51
Scale-out workload solution
    analysis and recommendations 39
    application architecture 37
    deployment 39–40
    designs 29
    experts and skills 36
    hardware inventory 33–35
    HPE Apollo-based solution 29–31
    management tools 40–41
    Processor Options 32
    project plan 35–36
    single-threaded application 31–32
    test application 37–38
    test workload data/results 38–39
Scality® software
    connector servers 47–48
    deployment 51–52
    design 43–44
    erasure coding 49–50

key-value store 44–46
licenses 48
scale-out 50–51
sizing calculations 49
storage servers 46–47
supervisor server 48
system management 51

SmartArray controllers 12, 66–67
Smart Update Manager (SUM) 200
Software Defined Storage (SDS) 240
Software inventory 132
BLc7000 Blade Infrastructure 99–100
business intelligence solution 16–19
Citrix XenApp 7.6 environment 109
ConvergedSystem 700 121–125
Disaster Recovery 186
Epic solution 121–125
eVDI 99–100
HC250 and backup 175
Moonshot 1500 chassis 79
SQL Server workload 150–151
virtual machines 163

Solid State Disk (SSD) 184
Solid State Drives (SSD) 119–120
SQL Server scale-up workload
analysis 156
application architecture 153
application testing 153–155
business requirements 145
characteristics 145
expertise and skills 152
functions 147
hardware inventory 149–150
implementation plan 151–152
Intl Xeon E7-v2 processors 148
legacy architecture 147–148

software inventory 150–151
Superdome X configuration 146–147
Superdome X deployment 148–149
technical requirements 146

SQL Server storage solution 149
Stackato 216–217
Statement of Work (SOW) 133
Storage Area Network (SAN) 157
Storage Attached Network (SAN) 240
StoreFront Analytics 114
StoreOnce device 171
StoreOnce-Veeam® solution
backup performance numbers 160, 161
characteristics 160
components and processes 158, 160
hardware inventory 161–163
implementation
  fiber channel instructions 164–165
  iSCSI instructions 165–167
rack diagram 158, 159
RPO and RTO 160
software inventory 163

StoreVirtual technology 105
StoreVirtual VSA
hypervisor requirements 85
performance and capacity 85–86

Subject Matter Expert (SME) 89–90
Superdome X platform
analysis 156
application architecture 153
application testing 153–155
configuration 146
deployment 148–149
expertise and skills 152
hardware inventory 149–150

implementation plan 151–152
Intl Xeon E7-v2 processors 148
NUMA 153–155
rack diagram 146–147
SAP HANA 137
scalability 146
software inventory 150–151

SUSE Linux Enterprise Server (SLES) 137

Synergy
advantages 239–240
components 240–241
management of 242

System-on-chip (SoC) processing 236

Systems Insight Manager (SIM) 40

T

Tailored Data Center Integration (TDI) 137
3PAR 7440 117
3PAR 7440 Hybrid Arrays 182
3Par Storage solution 97
3PAR StoreServ 7440c 4N storage solution 113
3PAR StoreServ software 114
Top of Rack (TOR) switches 96, 128
Total cost of ownership (TCO) 73
2x D6020 60
2x DL380 Gen 9 servers 60

U

Universal Extensible Firmware Interface (UEFI) 76–77
Universal memory 236

V

Veeam® software
backup performance numbers 160, 161
characteristics 160

components and processes 158, 160
hardware inventory 161–163
implementation
fiber channel instructions 164–165
iSCSI instructions 165–167
rack diagram 158, 159
RPO and RTO 160
software inventory 163

Virtual CPUs (vCPU) 83, 85, 134
Virtual Delivery Agent (VDA) 107
Virtual Desktop Infrastructure (VDI)
customer requirements 81
processor selection 84–85
solution inventory 86–88
solution overview 81–83
StoreVirtual VSA
hypervisor requirements 85
performance and capacity 85–86
user activity assessment 83
vCPU 83
from Windows 7 to Windows 10 83

Virtualization Performance Viewer 114, 121

Virtual machines (VMs)
HC250
goals 169
implementation, deployment, and testing 176
IOPs 170
mega-bytes per second 170–171
parameters 171
software inventory 175
solution validation and criteria 176
StoreOnce device 171
test workload results 176–177
StoreOnce-Veeam® solution
backup performance numbers 160, 161

characteristics 160
components and processes 158, 160
hardware inventory 161–163
implementation
    fiber channel instructions 164–165
    iSCSI instructions 157, 165–167
rack diagram 158, 159
RPO and RTO 160
software inventory 163

VM Vending Machine 211

VMware Horizon 96

vSphere Metro Storage Cluster (vMSC) 105

W

Windows Operating System Server Version 106

Workload and density optimized (WDO) servers 193

Workloads
categorization 4–6
evolution 2–4
framework 7–9
IoT 6–7

X

XL170r Gen 9 129